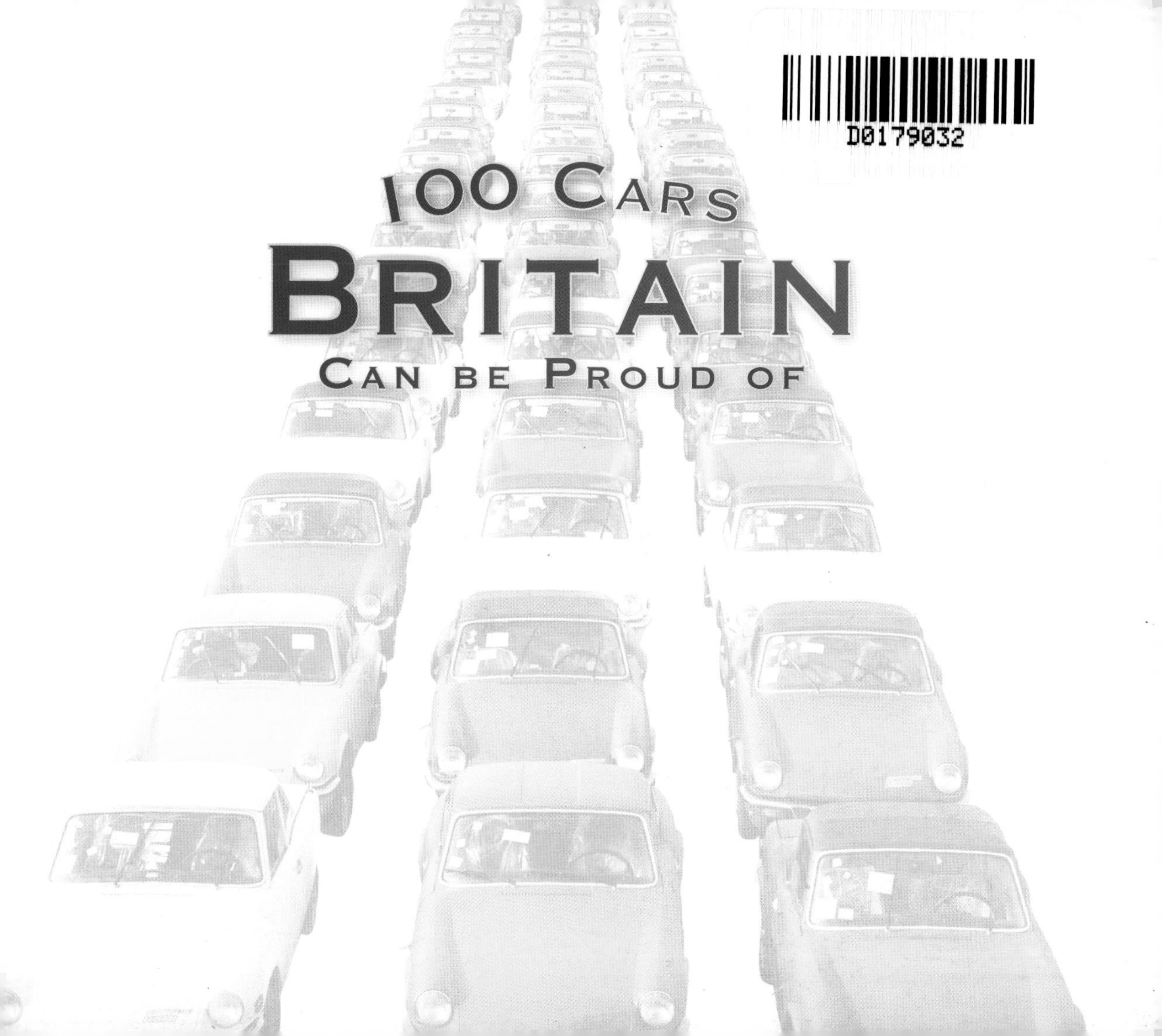

100 Cars Britain Can be Proud of

100 Cars Britain Can be Proud of

Giles Chapman

For Freddy, always so full of encouragement

First published 2010

The History Press
The Mill, Brimscombe Port
Stroud, Gloucestershire, GL5 2QG
www.thehistorypress.co.uk

British Library Cataloguing in Publication Data.
A catalogue record for this book is available from the British Library.

ISBN 978 0 7524 5686 7

Typesetting and origination by The History Press
Printed in India

INTRODUCTION

GREAT BRITAIN has one of the world's truly astounding records for producing exciting, ground-breaking and popular cars.

This island nation has given life to more great marques per head of driver population than anywhere else, with names that conjure up the most distinctive vehicles on the road. There is still nothing else with comparable character and heritage to match an Aston Martin, Bentley, Bristol, Jaguar, Lotus, Land Rover, Mini, Morgan or, of course, a Rolls-Royce.

Even the cars we *don't* make any more – such as Jensens, Morrises, Rovers, Triumphs and TVRs – inspire extraordinary desire and affection worldwide.

But we Brits do have a bewildering tendency to think our motor industry has gone to the dogs, and that the 'Great British Motor Car' is a thing of the increasingly distant past.

This book sets out to redress the balance. In it you'll find 100 cars that Britain should be resolutely proud to celebrate – 100 stories of vividly contrasting road-going machines that will blip the throttle of the sourest car curmudgeon. Not just those magnificent pioneering, vintage and classic machines of the glorious past, either, but also the new cars that, today, are keeping British designers, engineers and car factories humming with activity.

THE AC ACE

The Ace is one of the most beautiful sports cars this country's produced, and offers a wonderful driving experience. If you were a well-heeled young buck in the 1950s, it was the one for you, provided you could tolerate the scowling envy from every man you sped past.

This AC (it stands for Auto Carriers – making three-wheeled delivery trucks was the firm's founding activity) has a convoluted background. The company forged its reputation with superb sports cars in the 1930s, but its 1946 2-litre saloon was pretty tame. Something new was desperately required when along came John Tojeiro, a small-time racing car constructor.

Tojeiro had designed an excellent ladder-frame chassis featuring all-round independent suspension. Although influenced by racing expertise, this one was intended for a two-seater sports car. 'Toj' had already done all the scientific stuff, but he'd never manufactured road cars, so AC was perfectly placed to buy the whole project from him.

HOW MANY ROASTIES?

The Hurlock family, with income from their haulage business, bought AC Cars in 1930, at the height of the Depression, and owned it until 1986. Its main business after the Second World War was making invalid carriages for the NHS! Visitors would often be asked to join Derek Hurlock for lunch at AC's Thames Ditton factory in Surrey, where the avuncular boss would ceremonially carve the joint in the dining room. . .

The Ace's delicate, Italian-inspired looks encased a capable sports-racing car.

Facts & Figures: The 'AC' Cobra

The Cobra, although based on the Ace, doesn't count as British. Texan millionaire Carroll Shelby created it by installing American Ford V8 engines in Ace bodies shipped over to California. It turned the refined English sports car into a flame-snorting American hot-rod that could wallop any Ferrari. About 1,000 were made but, because impecunious enthusiasts coveted them so desperately, it's said 100 times that number of replicas exist, making the Cobra the most copied car ever.

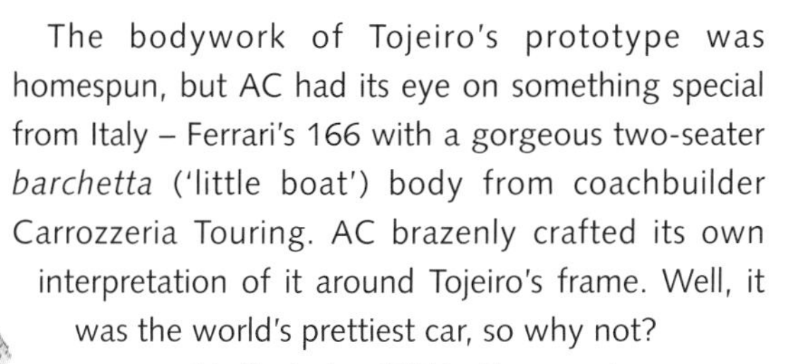

The bodywork of Tojeiro's prototype was homespun, but AC had its eye on something special from Italy – Ferrari's 166 with a gorgeous two-seater *barchetta* ('little boat') body from coachbuilder Carrozzeria Touring. AC brazenly crafted its own interpretation of it around Tojeiro's frame. Well, it was the world's prettiest car, so why not?

At first, in 1954, the engine was a sluggard. AC used its own straight-six 2-litre, born in 1919, giving just 80bhp (tickled up to 102bhp by 1958) for a fairly sedate performance. Bristol's 130bhp 2-litre firecracker was offered from 1956. Top speed rocketed to 115mph and 0–60mph now took 9 seconds. When Bristol stopped making it in 1961, AC adopted Ford's Zephyr 'six', often upgraded to triple-carburettor form by Ken Rudd, himself a formidable race competitor in his Ace-Bristol.

The exclusive Ace (only 723 were built) was highly prolific and successful in British production sports car racing, and robust enough to finish seventh overall – and first in its class – at the 1959 Le Mans 24-hour race.

▲ *Here's a later Ace, with 2.6-litre, triple-carburettor Ford Zephyr power.*

◄ *AC-built Cobras off to California to receive their V8 hearts.*

THE ALVIS TD21

Alvis is a marque adored by a tight-knit coterie of devotees but one that, for outsiders, is quite tricky to get a handle on. They're classic sporting cars in a hinterland between Jaguar and Aston Martin; they're elegant tourers, but not quite as splendid as Bentleys.

In fact, the TD21 and its derivatives carried Alvis gently towards the conclusion of its car-making period in 1967; after that, it concentrated on military vehicles but, in truth, cars increasingly had been a sideline since the 1930s, when the company discovered how much dosh could be made from turning out aircraft engines.

Alvises were motor cars for mature, discerning enthusiasts – luxurious, well-built, and yet with a surprising turn of speed. The TD was a steady evolution of the previous TC21 cars and one of these, called the Grey Lady, could muster 100mph from its 100bhp 3-litre straight-six engine. It had an old-fashioned ambience at odds with its energetic performance.

A 1959 TD21, conservative but beautifully engineered.

The Alvis TE & TF21

The 1963 TE21 featured four headlamps, and could be had with automatic transmission and power steering, and the 1965 TF21 was the last-of-the-line – with a triple-carb, 150bhp motor allowing 120mph performance. *Autocar* magazine summed up the cars perfectly in 1965 with this tweedy verdict on the TE: '. . . Combines pleasing qualities of the past with desirable features of the moment. It retains the Alvis character, more than a touch of traditional craftsmanship, and is clearly a smart turnout.'

▲ *Artwork from a TE21 brochure encapsulating Alvis's thoroughbred nature.*

As there was no-one at Alvis artistic enough to come up with a modern-looking body, the company adopted a Swiss design by Hermann Graber, and this became the elegant TD21 in 1958. It could be ordered as a roomy four-seater saloon or convertible; the comfortable interior was richly upholstered in well-padded leather, with the occupants' radiant expressions reflected in a gleaming walnut dashboard.

It was now a 110mph car too, thanks to an engine giving 120bhp, and the Alvis engineering department made valiant attempts to pack in the latest kit: in August 1962 came disc brakes, and three months later a five-speed ZF gearbox, as used by Aston Martin, was fitted.

STILL GOING STRONG

Alvis, you may be amazed to know, still exists in all but name. Red Triangle Autoservices took on the owner support role from Alvis in 1968, including all the spares and detailed records relating to every one of the 21,250 Alvis cars built in Coventry. All of which is extremely comforting to the owners of the 4,397 surviving examples.

However, these handsome, handmade machines still had a vintage feel – something 'young fogeys', viewing the anything-goes 1960s with mild horror, clung on to for dear life. As for famous owners, the only one brash enough to boast about Alvis TD21 ownership has been Nicholas Parsons – about as far removed from Mick Jagger as you can get.

THE ASTON MARTIN DB2

Yorkshire engineering tycoon David Brown was a lifelong speed freak. And a bargain-spotter to boot. He was browsing *The Times* one day in 1947 when he spotted an advert offering a sports car maker for sale for £20,000: it was Aston Martin, and he bought it. Soon afterwards, he picked up Lagonda (price: £55,000) because *it* had a brand new six-cylinder, twin overhead-camshaft engine on the stocks, designed by none other than the legendary W.O. Bentley.

This was ripe for marriage to Aston's promising Atom chassis but, first, prototype Atom-based coupés contested the 1949 Spa and Le Mans 24-hour endurance races. Both Aston and Lagonda engines were run, but the 2.6-litre Lagonda proved itself infinitely the superior.

DB2s storm Le Mans

Two Aston Martin DB2 Vantages entered Le Mans in 1950 and, as production models, they were even driven to the circuit. They ran like trains in the event. George Abecassis and Lance Macklin brought theirs home fifth overall, while Reg Parnell and Charlie Brackenbury took sixth. For 1951, factory team cars finished third, fifth and seventh, while privately entered DB2s came tenth and eleventh. Five finishes from five starters in this 24-hour killer was absolutely unheard of.

In fact, the Aston Martin DB2 that went on sale in 1950 at £1,920 was derived directly from this lightweight experimental racer, with small revisions. Customers could opt for coupé or convertible, two-seaters panelled in aluminium, and a gearlever mounted on the floor or steering column. Suspension was by coil springs and trailing arms, with the live rear axle

This ex-Le Mans DB2 fetched £500,000 at auction in 2009; it finished fifth in 1950.

THE ASTON AND MARTIN OF
ASTON MARTIN

Aston Martin was born in a Kensington garage in 1921 – and backed with Lionel Martin's family fortune from the English China Clay pits in Cornwall. The 'Aston' part of the car's name derives from a Buckinghamshire hill at Aston Clinton. It was a regular venue for challenging uphill races, and there's a cairn stone memorial there today commemorating the link between car and promontory.

located by Panhard rod, a system unaltered during the car's production life.

The DB2 betrayed its origins as a competition machine. The coupé's tiny rear window gave rotten visibility and the luggage area could be loaded only from inside – there was no boot lid. Then again, the entire frontage could be swung open for unimpeded mechanical access, which the DB2 also owed to its pit-lane origins.

In standard form the DB2 had 105bhp, and promised 116mph and 0–60mph in 11.2 seconds – sizzling stuff for the times. An optional 125bhp 'Vantage' engine made the DB2 beefier still. From appealing mainly to motor racing anoraks, Aston Martin was now the very first purveyor of a race-bred GT car, mixing roadholding, style, refinement, craftsmanship and sheer British Bulldog character. David Brown was chuffed, while monied connoisseurs naturally craved DB2s of their own, and 410 were built.

David Brown, on the left, delivers an aptly-registered DB2 to its new owner, Lord Brabazon.

The lines were extremely sleek for 1950, as the DB2 was a race-bred car.

THE ASTON MARTIN DB5

For some car-mad picturegoers, the star turn in *Goldfinger*, James Bond's third big movie adventure in 1964, wasn't Sean Connery at all, but something with equally hairy-chested sex-appeal: a silver Aston Martin DB5. Bond's most impressive on-screen gadget by a country mile, the Aston had film fans agog as an ejector seat sent his enemy hurtling through the roof. Replicas of the *Goldfinger* DB5 toured the world, drawing huge crowds years after the film had left cinemas.

But what of the DB5 in real life? Well, for starters, you had to be in the playboy demographic to so much as swing open one of its lovingly crafted doors. When launched in 1963 it cost £4,249, when you could get an E-type for less than half that. Even with the tuned, and optional, 314bhp Vantage engine, it could only just out-run the Jag to reach 150mph.

THE ROYAL BLESSING

Prince Charles was a little young for the DB5. His very first car was an MGC GT, but on his 21st birthday the queen and the Duke of Edinburgh presented him with an Aston Martin DB6 Volante. He owns it to this day and, true to his eco-friendly ethos, has even had the car converted to run on environmentally friendly bio-fuel, including surplus wine. . .

◄ *Sheer GT class.* ▼ *DB5 (on left) and DB6 shooting brakes.*

Facts & Figures: Surely the ultimate estate car?

By the time the DB5 was introduced, Aston Martin had made its home in Newport Pagnell, Buckinghamshire. That's where all the 1,063 examples of the car hailed from . . . although 12 of them were completed as super-cool shooting brakes at a London workshop. Some of these were bought by company owner David Brown's shootin' and fishin' social circle, so there was immense cachet in owning one of the rarest DB5s of all.

However, being entirely handmade in aluminium over a tubular steel frame, the DB5 was every bit as urbane as Ian Fleming desired. The DB5 was a subtle but thorough update of the 3.7-litre DB4 first seen in 1958. It was instantly recognisable with its nose boasting sleek, faired-in headlights. With a bigger, 4-litre version of the twin-camshaft, six-cylinder engine offering, as standard, 240bhp, plus a five-speed ZF gearbox, Aston Martin could provide more power, torque and cruising ability.

A relaxed Sean Connery poses with 007's DB5.

'It is a car requiring skill and muscle – a man's car – which challenges and satisfies and always excites,' salivated *Autocar* in a 1964 road test, having just accelerated one up to 140mph and then brought it to a reassuring halt thanks to the Aston's four-wheel disc brakes.

Yet the DB5 actually lagged behind in technical innovation. Independent rear suspension was eschewed for a traditional solid back axle, which made the DB5 better suited to long, fast highways than twisty mountain roads. Inside, meanwhile, the atmosphere was proudly old-fashioned, with the leather upholstery and trim redolent of a London gentleman's club, rather than a mere motor vehicle.

But that's not to suggest Bond-style gadgetry was absent. Electric windows were standard (and uncommon touches in 1963), and when the push-button radio was switched on, the words 'Aston Martin' glowed in red from the tuner. Who couldn't relish seeing that each morning?

THE ASTON MARTIN V8

This muscular British supercar was to dominate everything Aston Martin did for the two turbulent decades of the 1970s and '80s. It became a rock-solid fixture of the motoring scene and – like a good port – aged slowly and steadily, so that its desirability was never really diminished despite its advancing years. In short, a matured-in-oak classic.

Those swoopy lines were first revealed as long ago as 1967 on a car called the DBS. The old 4-litre straight-six power unit was carried over from the DB6, but the DBS was streets ahead – literally – in the handling stakes, because a De Dion independent rear suspension system transformed the handling from unwieldy to nimble, for its bulk.

But there was a real treat in store for 1969. Aston had been carefully perfecting its 5.3-litre V8 engine for yonks, and now here it was under the DBS's wedge-shaped bonnet; all 375bhp of quad-camshaft, all-

THE LIVING DAYLIGHTS

Like an ageing matinée idol enjoying a late-career fillip, the V8 finally appeared in a James Bond movie in *The Living Daylights* in 1987. The special effects chaps fitted it with outrigger skis to help it tackle one snowy scene. The film, indeed, needed every bit of available extra excitement, to make up for Timothy Dalton's panache-free portrayal of 007.

The V8 remained the beefy British alternative to Italian supercars for two decades.

aluminium, hand-assembled motor just awaiting the command – via the driver's right foot – to catapult the monster to 160mph.

In this basic form, the car would serve Aston Martin through its many ups and downs until 1990. It was often categorised as a lovable anachronism, although the Virage that replaced it was no less of a dinosaur. V8 career highlights numbered the replacement of fuel-injection by carburettors in 1973 (only for injection to reappear 13 years later), the delectable Volante convertible introduced in 1978, and the super-fast V8 Vantage.

Financier David Brown decided he'd had enough fun with Aston Martin and sold up in 1972. The company's survival after that, in retrospect, now seems an absolute miracle, as it coincided with a worldwide oil crisis and a sustained period of rampant inflation. In the end, it fell to ebullient Aston fanatic Victor Gauntlett to keep the Aston Martin ship afloat, a task he tackled with rousing British determination.

Facts & Figures

Services rendered at Newport Pagnell

Between 1926 and 1957, Aston Martin was one institution that Feltham, Middlesex, could be proud of. But the move to Tickford Street, Newport Pagnell, in '57 made sense, as it united the engineering and the coachbuilding activities. Thanks to its M1 associations, Newport Pagnell hardly oozes romance, but Aston Martin put the Buckinghamshire town on the map. Although carmaking had shifted entirely to Warwickshire by 2007, the historic site remains the company's service and restoration hub.

THE ASTON MARTIN V12 VANTAGE

If Mr Jeremy Clarkson is to be believed, then the V12 Vantage will be one of the last truly great cars ever offered to the public. In a memorable finale to the autumn 2009 series of BBC2's *Top Gear*, the cameras swept overhead as he drove the car across stunning Scottish vistas; in place of Clarkson's usual heavy intonations, came dreamy, soothing music. We were invited to simply drink it all in, and his verdict was simple: 'It's wonderful, wonderful, wonderful.'

Top Gear has a deep love for Aston Martins, but Clarkson suspects bureaucrats and environmentalists will conspire to legislate cars like the V12 Vantage out of existence.

Facts & Figures: The Aston Martin V8 Vantage

In 2006, Aston Martin decided finally to turn its fire on Porsche, with a small, agile junior supercar offering a distinctively British alternative to the wearyingly accomplished 911. The V8 Vantage was an Aston watershed, with its tiny dimensions and wholly purposeful design – even a highly practical hatchback. Just to rub it in, the 4.3-litre V8 engine was assembled in Germany, and the new six-speed baby Aston was handbuilt at a slick new factory near Warwick. Critics and customers absolutely loved it.

If you're unmoved by the gorgeousness of the V12 Vantage then the nanny state has won!

After all, it's a machine built for little else than unalloyed driving exhilaration, reserved for a wealthy élite. Using all the mouth-watering assets at its disposal, Aston Martin slotted the 510bhp V12 engine from its luscious DBS into the compact V8 Vantage, a two-door, two-seater coupé with a bonded aluminium structure that makes it an elf beside the ogres of traditional Astons, such as the 1980s V8 and the Virage.

This cocktail, weighing in at just 1,680kg thanks to additions like a carbon fibre bonnet and boot, makes this a 190mph car that can sprint to 60mph from standstill in 4.1 seconds. Even the upholstery is lightweight, which means Alcantara fabric rather than leather, and minimal carbon fibre seats are available as an option (for performance obsessives disciplined enough to resist pork pies).

Glueing it to the road are a rear airflow diffuser and an adjustable rear aerofoil. Multiple bonnet louvres ventilate that amazing 6-litre engine and carbon-ceramic brakes perform the manful duty of bringing this ultimate Vantage to a stop.

Despite its crushing performance credentials, the V12 Vantage isn't a racing car. Its only reason to exist is to delight the lucky few with £135,000 to spend on the most thrilling Aston of modern times. And there can be no more than 500 a year, the company insists. You have to admit, it's just begging to be swamped by Britain's killjoy army. . .

BACK IN BRITISH HANDS

Between 1987 and 2007, Aston Martin was owned by Ford. Henry Ford II was apparently persuaded to buy the company as a trophy asset by his wife, and by Walter Hayes, Ford of Europe chairman, because of its glittering image. In 2007, it changed hands for £475m, as a British-led (but Kuwaiti-backed) consortium stepped in. David Richards, founder of rally team Prodrive, is now in charge.

➤ *It's quite practical, you know: 190mph and a Waitrose-friendly hatchback.*

THE AUSTIN SEVEN

Herbert Austin bade farewell to his former employers at Wolseley in 1905 to establish his own car company; by the early 1920s it was one of the biggest in Britain, with a huge payroll of 22,000 workers swelled through big wartime contracts. But when the demand for army trucks and ambulances halted, the Austin Motor Company hit hard times. This was not helped by the government's 1920 Motor Car Act, which clobbered large-engined cars like Austin's regal 20hp with tax penalties that many potential customers found unpalatable. Turnover fell and 'The Austin' faced insolvency as the company briefly operated under administrators.

Clearly a 'big idea' was required, but salvation actually came along as an idea for something little. Herbert Austin decided that he was going to 'motorise the common man' with a very small, very economical car. He may have been motivated by the many cheap-to-buy 'cyclecars' offered by enterprising French companies, but this type of flimsy device was far from his mind.

Facts & Figures: Britain's first 'nice little runner'

The launch price of the Seven was £165 but, as production was ramped up, that soon fell. Because 302,000 were made up to 1939, used Austin Sevens virtually created the second-hand car market in Britain. This means new and used Sevens introduced more middle and working class Britons to the convenience and pleasure of motoring than probably any other car before the Second World War.

➤ *The 1928 original.*

◄ *A 1932 Seven in Arrow foursome sports guise.*

He intended to make a real car but in miniature, with none of the compromises of using motorcycle parts. Still, it would have to sell for around the same money as a motorbike/sidecar combination.

Austin closeted himself away in the billiard room of his house with a talented 18-year-old draughtsman called Stanley Edge, and together they created what became the Austin Seven. 'He [Herbert Austin] was hellishly bad-tempered,' Edge recalled, decades afterwards. 'I only learned later that he was having quite a battle with his directors to get money allocated for the Seven.'

Unveiled in 1922, the Austin Seven offered four seats, brakes on all four wheels, and 40mph from its proper four-cylinder 747cc engine. Despite sarcastic jibes about 'needing one for each foot' thanks to its tiny 9ft length, the Seven became unquestionably the planet's leading small car. Rights were sold worldwide to build it abroad: BMW's first cars were licence-built Austin Sevens, while other versions were popular in France and even the USA.

The Seven put Britain on wheels.

That dastardly Datsun doppelganger.

GIVING JAPAN A BAD NAME

No-one from Japan will ever admit it but the very first Datsun, the 1932 Type 10, was a blatant rip-off of the Austin Seven, albeit with an even smaller four-cylinder motor of just 495cc. It helped give Japan's carmakers a dodgy reputation for copying Western technology that took 60 years to throw off.

THE AUSTIN A30/A35

The A30 and A35 are reliable little cars with a big reputation for staying power. Among economy runabouts from the 1950s, they are outdone only by the Morris Minor for their ability to last and last. That's the reason you still see quite a few of these upright little machines – the garden gnomes of the classic car world – on Britain's roads.

This must surely have been closely related to the fact that the A30, which was introduced in 1951, was the very first Austin without a separate chassis. It was the vogue at the time for carmakers to switch to monocoque construction, with the body and chassis as a single, all-welded structure, which made manufacturing easier and would begin to produce more refined cars that shuddered less and were quieter to use.

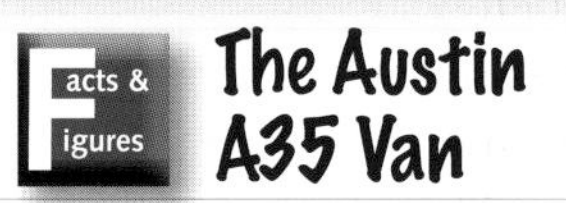

The Austin A35 Van

If Austin products could still find customers then the factory was generally loath to banish them from its production lines. British businesses large and small valued the A35 van, and so it continued to be available until 1968, by which time it was a living antique. It had its admirers in the fast set, too; James Hunt (pictured below), Formula 1 World Champion in 1976, ended his days virtually penniless, yet still maintained he got great enjoyment from the A35 van he drove (it had been his mum's) until his death in 1993.

A two-door A30, photographed in 1953.

So no doubt Austin did the job extra carefully, over-engineering the inner parts of the car so it wouldn't flex like an old tin bath. At the same time, the mini-battleship approach meant it could stand up stubbornly to the onset of rust.

The A30 had its shortcomings, especially in its old-fashioned mechanical brakes and a narrow track that made it feel like it would topple over in blustery side-winds. On the other hand, with just 803cc under the bonnet, you were never going to be stopped for speeding. Just weaving!

There was some added zing in 1956 with the arrival of the A35. It was the same car but with a 948cc engine, a bigger back window, smaller wheels to lower the ride height and higher gearing to make up for them.

Although the A30 had been planned as a Morris Minor rival, the cars were now stablemates in the British Motor Corporation. In fact, they had the very same engine and gearbox. Yet despite the A30's pedestrian image, it was a rather better car than competitors from Ford or Standard. In fact, it was still winning friends – and sometimes even the odd saloon car race – when the Mini replaced the A35 in 1959.

WHAT A BARGAIN!

The first Austin A30, a four-door saloon, arrived in 1951 with a price tag that was sorely tempting. It cost just £520. Morris must have been absolutely mortified: it was asking £632 for its cheapest four-door Minor – a whopping 22 per cent premium! Within a year, the former enemies had merged. . .

➤ *The compact A30 was the first Austin to do away with a separate chassis.*

THE AUSTIN & MORRIS 1100

Following the acclaim garlanded on the Mini in 1959, the British Motor Corporation was counting on its technical director, Alec Issigonis, to repeat the achievement. Three years later, therefore, a bigger BMC family saloon would share the Mini's strengths of transverse engine, front-wheel drive, subframe mechanical systems, and a roomy interior.

But it was actually an even more sophisticated car. It boasted an interconnected fluid suspension system called 'Hydrolastic', designed by Dr Alex Moulton, which gave a remarkably smooth ride. Under the bonnet sat a 1,098cc version of the BMC A-Series engine, front disc brakes were standard, and automatic transmission was available, while detail styling polish was handled by Italy's Pininfarina studios.

First on the market in August 1962 was the Morris 1100, but other derivatives appeared thick and fast. In fact, BMC's 1100 became an intensively 'badge-engineered' product, with seven separate guises: Austin, Innocenti, Morris, Riley, Wolseley, MG and Vanden Plas. Each had a unique livery, and upmarket models had twin carbs for an extra 7bhp. There were two- or four-door saloons and, from 1966, Austin Countryman or Morris Traveller three-door estates.

MkII models arrived in 1967, heralding a 58bhp 1,275cc engine option. Added enthusiast-appeal

▲ *The Morris 1100 gave the British family car a high-tech edge in 1962.*

THE AUSTIN ALLEGRO

The Allegro that supplanted the 1100/1300 in 1973 as British Leyland's staple small family car never made the nation proud when it was current. It was a constant reminder of everything that was wrong with that organisation's attitudes to the motoring public, from misguided design to frustratingly uneven quality. At 667,000 sales in 10 years, it never remotely matched its predecessor's popularity. Nor, though, was it as dire as many claim . . . and its kitsch kudos today is immense!

Facts & Figures: Much Ado About Something

The secret codename of the BMC 1100 during its chrysalis period was ADO16 – the 16th project to emerge from the Austin Drawing Office. This heavily guarded area was part of Austin's gigantic Longbridge factory a few miles south of Birmingham, which was often called 'The Kremlin' by lowly employees.

▲ *This mock-up, built in 1959 at Longbridge, paved the way for ADO 16.*

arrived in 1969 with the four-door 1300GT, badged Austin or Morris, boasting a 70bhp twin-carb motor good for 96mph.

These small saloons were lapped up by us Brits. In 1965/6 and again from 1968–71, this was the nation's best-selling model range, gamely holding the Ford Cortina off the top spot. Over two million were shifted in total. We enjoyed their crisp style, superior handling and spaciousness. In fact, the only person who seemed to dislike the car was John Cleese, who famously gave his Austin 1100 Countryman a 'damn good thrashing' in *Fawlty Towers*.

It's easy to overlook this ground-breaking car today and, shamefully, its most avid following is in Japan, where the chrome, leather and walnut-laden Vanden Plas 1100 is considered motoring ambrosia.

THE AUSTIN METRO

Wonder if you've forgotten the slightly bizarre official title of Britain's answer to those foreign superminis, on the day it was revealed – Friday, 8 October 1980?

British Leyland's boffins had called it LC8 while developing it in secret, but a 1979 poll of BL workers put the 'Metro' name streets ahead of other front-runners like Match and Maestro. However, Birmingham's train manufacturer Metro-Cammell got wind of the choice, and insisted BL could only use the Metro name if prefixed with the 'mini' tag. Presumably, they didn't want customers getting confused between the new three-door hatchback and a Piccadilly Line tube train. Still, both sides soon forgot about the agreement and the newcomer became the Austin Metro we all now know.

The Metro in its original form, 'A British Car To Beat The World'.

Facts & Figures: Rough but ready

One motoring journalist was given the exclusive opportunity to thrash an early 998cc version around Europe. He found the engine excessively noisy, the gearbox emitting a strange whining sound, and the entire power unit getting the uncontrollable shakes at 5,000rpm. Still, he declared it 'this tremendously important and interesting machine.' What a patriot!

The little car made a huge impact because it had been conceived against the depressing backdrop of British Leyland's grimmest years. So the rapturous reception it received was a rare bright spot among the prevailing talk of strikes, factory closures, and anti-union tirades from new prime minister Margaret Thatcher.

A few aspects of the Metro equalled, or even bettered, the Ford Fiesta, Renault 5 and Volkswagen Polo, particularly its interior space and nimble road manners. Several elements, though, were crummy because the entire drivetrain was borrowed from the 21-year-old Mini, giving an awkward, milk float-like driving position, a four-speed gearbox, and a distinct lack of mechanical refinement.

On TV over the first few weeks of the Metro's official existence, commercials showed a regiment of Metros holding back European rivals from a vantage point above Dover's white cliffs. Was it over-hyped? Of course! But the British public seemed oblivious to the Metro's flaws; orders cascaded in. By June 1981 a newly-minted Metro trundled out of the Longbridge factory every minute. By the end of the year it was Britain's no. 4 bestseller at 110,283, and it remained in the top five until 1988. In 1983, along came the Peugeot 205 and Fiat Uno. Both totally overshadowed the Austin in virtually every respect. So amazing, then, to think the Metro continued to be made, much modified but with many of its original downsides resolutely intact, until 1998.

THE COMEBACK KID

Fully 10 years after the country went mad for the Metro, Rover gave it a full going-over. There were refined, brand new, all-aluminium K-Series engines, slick five-speed gearboxes, and interconnected Hydragas suspension that gave the little cars the best ride and handling in their class. Trouble was, the metal architecture encasing all this driving vitality was really showing its age. The original Metro's proudest boast was its interior space; now it felt decidedly cramped.

TV hosts Terry Wogan and Sue Cook get a kick out of a Metro milestone.

THE AUSTIN-HEALEY 3000

At £1,064 in 1953, the Austin-Healey 100 became the world's cheapest 100mph sports car, and certainly among the most beautiful. With a 90bhp, four-cylinder, 2.6-litre Austin engine in a narrow chassis, acceleration was truly thrilling. The rakish 100, on wire wheels and with a fold-flat windscreen, could hit 60mph in under 10 seconds. From the start, the cars were very rare in the UK, with 90 per cent exported, overwhelmingly to the USA.

In 1956 it was replaced by the £1,144 100-Six with a straight-six, 2.6-litre motor providing 102bhp; two tiny rear children's seats were fitted and an oval, chrome-lipped grille superseded the 100/4's distinctive fan-shaped one. Despite the power hike, the 100-Six took 2 seconds longer than its predecessor to reach 60mph, but another power boost later restored the sparkle, and a two-seater was hastily reintroduced for those young blades not in the family way.

The car was renamed the 3000 in 1959 when it received a larger 2.9-litre engine giving 124bhp (and ultimately 148bhp). Front disc brakes were standard on this £1,159 roadster, which now gained the 'Big

The 'Big Healey' in 3000 Mk II guise.

WILL THE
BIG HEALEY RIDE AGAIN?

In 2006, Austin-Healey fanatic Tim Fenna gained permission from the Healey family to launch a new Healey sports car in the spirit of the original Austin-Healey 3000. With American backing, plans were announced for a two-seater roadster but, with the world in deep economic mire, it's all gone a bit quiet. Still, the considerable pressure of creating a new Healey car that doesn't suffer by comparison to the original has been avoided. . .

Healey' nickname for the manly growler under that long, long bonnet. Americans began to regard it as more poser than sprinter, because the 3000 faced stiff opposition from the home-grown V8 Corvette. Nevertheless, the BMC Competitions Department embarked on a competition programme with the Big Healey, and it proved remarkably tough and capable on gruelling cross-country rallies.

A 3000 won its class on the 1960 Alpine Rally, and then Pat Moss – Stirling's sister – gave the car its first outright win on the 1960 Liège–Rome–Liège Rally. The cars did well on the 1962 Monte Carlo and Alpine Rallies too, in the hands of the Morley brothers and Paddy Hopkirk.

American safety laws and BMC apathy killed off the Big Healey in 1967. By then it was a bit old-fashioned anyway: the suspension was spine-jarring and engine heat turned the cockpit into an uncomfortable sauna on summer days. But it still sounded fantastic! Today a Healey 3000 is perhaps the epitome of the classic British sports car.

Facts & Figures

On-the-spot deal

The Healey 100 was the undisputed star of the 1952 London Motor Show. Leonard Lord was particularly bowled over. Over a hastily arranged dinner on the eve of the show's opening, the bombastic Austin chief negotiated hard for marketing rights with entrepreneur Donald Healey; by the time the show closed, the low-slung, two-seater had become the Austin-Healey 100.

➤ *The more elaborate 3000 contrasts with the original 100/4 above.*

THE AUSTIN-HEALEY SPRITE

The Austin-Healey Sprite brought fun and fresh air to a whole generation of cash-strapped sports car addicts. During the 1950s, brand new sports cars were the preserve of the wealthy but, upon its arrival in 1958, the Sprite changed everything. It was easily the cheapest two-seater made by a proper manufacturer, and the fact it sold nearly 50,000 examples in three years says it all. The advanced construction followed the pattern of the Le Mans-winning Jaguar D-Type: a stiff central unitary structure with outriggers front and rear to carry the drivetrain and suspension. All panels were kept as simple as possible: the entire inner body structure was flat metal, and it was initially planned to make the front and rear bodywork absolutely identical. As it was, the dinky little body didn't look far off that concept anyway, with a one-piece bonnet/wings section opening up, hippopotamus-style, for unimpeded engine access.

➤ *Until the Sprite came along, budget sports car fun was an elusive commodity.*

There were sidescreens instead of windows, no external door handles, and no boot lid; luggage had to be stuffed in from behind the two wafer-thin seats.

As for a power unit, an Austin A35 948cc A-Series engine was equipped with twin SU carburettors to give 43bhp (a 9bhp power increase), while the A35's four-speed gearbox, front suspension and rear axle went in unaltered. However, not all A35 organs were conducive to seat-of-the-pants driving, so the rack-and-pinion steering came from the Morris Minor.

Austin launched the new sports car in 1958 at £660 including Purchase Tax. It was bang on. At that price, the Sprite simply had no peers: an MGA cost £995, a Triumph TR3 £1,050. Until the Triumph Spitfire appeared four years later, its principal rivals were horrible, home-build, plastic kit cars.

'Frogeye'

The Sprite was going to have-pop-up headlights until the idea was quashed on cost grounds, so instead Healey stuck the lamps on top of the bonnet, set close together and bug-like. There were contemporary grumblings at this kooky headlamp treatment – ironic since they made the Sprite one of the most instantly recognisable cars on the road, and quickly garnered the 'Frogeye' nickname it has never lost.

The Frogeye was no firebrand, with a top speed of just 83mph, but its meagre weight still allowed it to dance around many conventional sports saloons in the handling stakes; well, until the advent of the Mini Cooper, anyway.

A LONG, LONG LIFE

A heavily 'normalised' Sprite Mk II arrived in 1961 with entirely new front and rear bodywork, an opening boot lid, and a bigger 1,098cc engine. In the process, the original's perky charm evaporated; an MG-branded edition, the Midget, accompanied it. Sprites were made until 1971 (the last few labelled as Austins rather than Austin-Healeys following a petty row over royalty payments between Donald Healey and BMC successor British Leyland), but the Midget soldiered on until 1979.

➤ *Once the cheeky headlights had been replaced (this is a 1971 Sprite) some character was lost, but popularity remained buoyant.*

THE BENTLEY 3-LITRE

Vintage (that is, pre-1931) Bentleys occupy an illustrious pedestal in British motoring history. It seems incredible they were made for only 10 years. They were the most successful sports cars of their day and, if anything, *too* well made. Bentley, never the most economically viable of enterprises, lavished fastidious care on its output, managing to build a mere 1,619 3-litres.

The typical Bentley was a sports car but many were saloons. Like all the smartest manufacturers, Bentley didn't make bodywork: one ordered a chassis that was despatched to one's favoured coachbuilder to be 'tailored'. Frequently, a Bentley's second owner would have the original purchaser's coachwork replaced to his own satisfaction, like a new homeowner ripping out a perfectly good kitchen. . .

Walter Owen Bentley, universally called 'WO', founded Bentley Motors in 1919 and, later that year, showed his first design, the 3-litre. It caused a sensation with its overhead-camshaft engine using four valves and two sparkplugs per cylinder – a car for the road built around motor racing technology. Since Bentley was a perfectionist engineer, the first car wasn't actually delivered until September 1921. It was expensive

A SHORT BURST OF GLORY

Bentley's independence lasted until 1931, when Rolls-Royce took it over; the Cricklewood-based company was bankrupt after spending heavily on racing and perfecting the supercharged 4.5-litre model. 'WO' stayed on as a consultant but the subsequent Rolls-based cars he had a hand in lacked the Bulldog spirit of the first Bentleys, excellent though they were.

The unmistakeable face of the vintage 3-litre Bentley.

This particular 3-litre, one of the earliest, belonged to the author's dad.

A sporty, short-chassis 3-litre.

but already had a race-bred pedigree; the working prototype itself won a race at the Brooklands circuit.

The customer could choose from three lengths of wheelbase, from 108in to 130in. Saloon and even limousine bodies were made on the longer chassis, while the short-wheelbase model was often kitted out as a two-seater. Bentleys gained great renown, not because the 3-litre was spectacularly powerful or fast – 65bhp and 82mph weren't extraordinary, even in 1922 – but because of the way the cars performed. Their engines were slow-revving, flexible and strong, the four-speed gearbox was sweet, with well-chosen ratios and, despite high ground-clearance, handling was predictable.

One ran in America's 1922 Indianapolis 500 and came home 13th against out-and-out racing cars. That same year, Bentleys finished 2-4-6 in the Tourist Trophy. 'WO' himself, competing in his last race, was among the drivers.

Facts & Figures

Circuits and rounds

In 1923 a 3-litre covered the fourth highest distance in the inaugural Le Mans 24-Hour race and the following year one driven by the duo of Duff/Clement covered the greatest distance of all to win. They ran at Le Mans in 1925 and 1926, without success, but one survived a major crash in 1927 to win once more. It was driven by S.C.H. 'Sammy' Davis of *Autocar* magazine and a Dr J.D. Benjafield . . . who used the car for his patient rounds afterwards.

Le Mans 1924, with W.O. Bentley in the middle.

THE BENTLEY R-TYPE CONTINENTAL

Meet the Most Exciting Car In The World. It cannot be said Britain holds that accolade all the time but, in 1952 at least, there was no doubting it. The R-Type Continental was the fastest genuine four-seater car on earth, for it could surpass 120mph effortlessly.

It was also one of the most beautiful. Crewe's flagship was an owner-driver super-coupé, its flowing profile influenced by the 1948 Cadillac 62 Coupé and shaped in Rolls-Royce's Hucknall wind tunnel. The Bentley grille still stood proud and tall but those finned rear wings helped make the Continental reassuringly stable at high speed.

The refined 4,566cc engine breathed more freely than in the standard R-Type saloon thanks to a higher compression ratio and a big-bore exhaust. Gearing was higher too, to promote fuss-free,

OVERSHADOWED BY THE JET AGE

The late Christian Hueber, R-Type Continental owner (he covered 80,000 miles in his) and co-founder of the Continental Register, summed up the car's moment in the limelight: 'This was a stupidly expensive car when new. People bought it because it was the ***only*** way to travel to the Riviera, a car for real long-distance driving. However, by 1955, if you wanted a private rocket-ship you could get on a jet aircraft and, after that, Bentleys got bigger and heavier and spent most of their lives driving round town.'

With 200 of the 208 made surviving, the R-type Continental is the antithesis of the throwaway car.

A rare beauty for crossing continents before the jet age dawned.

high-speed cruising. Its alloy bodywork was built by H.J. Mulliner in west London, on a special high-performance chassis with a lowered scuttle and steering column. At 3,696lb, with the body weighing 750lb, 240lb was cut from the mass of a standard Bentley R-Type. The weight-loss regime for the first sporting Bentley since the 1930s included aluminium bumpers, not steel, and even lightweight alloy bucket seat frames.

Still, Continental occupants hardly had to rough it: four of them and all their luggage could travel in magnificent comfort in drawing room surroundings, the driver enjoying a full set of instruments including rev-counter and oil temperature gauge. On the road that tall gearing gave the Continental a fantastically long stride, with 80mph on tap in second, 100 in third and 120 in top.

Later Continentals diluted the original 'lightweight' R-Type concept. Portly tycoons demanded fatter seats, lazy drivers wanted automatic gearboxes. It became just another heavy, coachbuilt Bentley. But, back in 1952, there was no faster or more stylish way of escaping the grey skies of austere, joyless post-war Britain for the seriously rich, idle or otherwise.

Facts & Figures: Survival of the slickest

Seeing an R-Type Continental anywhere is something of an event, as a mere 208 examples were built between 1952 and 1955. But you're guaranteed never to have seen one in a scrapyard. The survival rate for these wonderful machines is unparalleled in the classic car world, with 200 still cherished by owners.

The S2 Continental of 1959 (here with Park Ward coachwork) was heavier and slower.

THE BENTLEY CONTINENTAL R

Could this be the finest all-British motor car made since the Second World War? You might well be able to pick fault with it next to today's Astons, Rollers, Bentleys and Jags in terms of gadgets; it might even be wanting somewhat in the ultimate power and top speed departments and, from an environmental viewpoint, it'll never be the least destructive.

But for sporting elegance, cosseting luxury and pleasingly old-school Britishness, the Continental R remains tough to better. Bentley had been trying to recapture some of its pre-war glamour throughout the 1980s. Mostly this was by the simple expedient of bolting a turbocharger to the Mulsanne saloon's colossal 6.75-litre V8 engine, to turn this road-going locomotive into a veritable bullet train. But then Bentley decided it really needed something unique in which to showcase this powerplant.

It brought in two keen young designers, Ken Greenley and John Heffernan, and told them to come up with a brand-new Continental. The pair worked in isolation in a disused west London factory.

Facts & Figures: Fit for a Sultan, and his pals

Bentley made 1,290 of these beautiful cars, with the very first one snapped up by the Sultan of Brunei with his customary disinterest in the actual purchase price of £199,750. He must have liked it very much because it became the starting point for some 600 special cars he ordered from Bentley, all of which were built in the utmost secrecy. They included weird variations like a Continental R four-door saloon and estate, many of which were given to family members and courtiers as gifts.

➤ Imposing and handsome, the Continental R was a return to form for Bentley in the 1990s.

The short-wheelbase Continental T is somewhat of a gentleman's dragster.

'When we'd finished we handed the final, full-size model over,' Heffernan recalled. 'And that was it – we didn't see the car again until 1990 when we were asked to come and test-drive it. I had trusted them completely and, apart from changing the headlights from rectangular to twin round ones, they had been completely faithful to our design. It was amazing what they'd done.'

Announced in 1991, the Bentley made a massive impact with its sophisticated looks, immense power and, of course, the most wonderfully sumptuous and finely crafted interior available anywhere. Heffernan recalled that Bentley anticipated selling relatively few cars but, in the event, 'it was so popular that the tooling was renewed twice!'

Offered only with automatic transmission and standing proud on its huge 18in wheels, the 150mph car was a sensation at the 1991 Geneva motor show. But precisely how powerful it was remained opaque; insider estimates suggested 325bhp and 450lb/ft of torque, delivered with a whispering roar. Bentley stuck to its discreet tradition of not revealing figures because it was deemed too vulgar, a spokesman once describing the turbo engine's power output as 'adequate, plus 50 per cent'.

PRESS FOR ZOOM SERVICE

Bentley made several attempts to sex-up the Continental R. Some, like the glass-topped SC Sedanca, were awful (controversial boxer, terribly non-U, Mike Tyson bought one), but others such as the short-chassis Continental T were gentlemen's dragsters with trick engines and pleasing touches like an engine-turned dashboard, chrome switches, and a red push-button starter for the engine.

THE BENTLEY MULSANNE

Unveiled at the 2009 Pebble Beach Concours d'Elegance, the Mulsanne stood out among the polished classics arrayed on the manicured Monterey sward. But traditionalists may have experienced that feeling of slight sinking once more; wasn't this yet more evidence of Bentley's treasured British heritage being craftily manipulated into being a 'luxury brand'?

Well, as it happens, this couldn't be further from the truth. The patrician Mulsanne is the most 'British' Bentley in ages, driving the super-luxury saloon concept on apace while sacrificing nothing that the old guard holds dear. For one thing, the engine may be pretty much all-new, all-aluminium and 15 per cent

SNAIL-PACED PERFECTION

Each Mulsanne takes 15 weeks to build, 170 hours of which is devoted to the interior, and at least 15 hours of that goes solely on hand-stitching the steering wheel. The body is spot-welded and brazed by hand. There are 114 paint colours to select from. There are 24 traditionally tanned leather choices, several different veneers and two marquetry styles. The finishing process for the stainless steel décor (totally ousting chrome – a world first) takes 10 hours. Of course, it can all take, and cost, an awful lot more if you ask Bentley's Mulliner division to bring your personal ideas to life.

▲ The Mulsanne's bodywork benefits from a huge amount of hand-crafting.

◄ The Mulsanne is the ultimate sporty limo.

► Pebble Beach launch alongside W.O. Bentley's personal 8-litre.

cleaner and less thirsty, but it sticks to the hallowed 6.75-litre capacity, with the V8 burble unharmed and the deep pools of pulling power unchanged, boasting 505bhp and 752lb/ft of torque. The automatic transmission is an eight-speeder.

Then there's the masterful design. Bentley owner Volkswagen has certainly parachuted in some foreign-sounding Johnnies to run Bentley but design director Dirk van Braekel, for one, has an uncommon understanding of Bentley's esteemed past. That's why the Mulsanne offers a marvellous blend of 1950s S-Type and Continental GT styles, without ever resorting to the slabbiness that sometimes afflicts, ahem, certain German cars. The long rear overhang is elegant, while the sculptural form of the front wings could never be pressed or handbeaten; they are possible only through cutting-edge aluminium superforming techniques.

Nor are all the mechanical and coachbuilding elements being surreptitiously sourced from China and Poland to cynically mine profit. The Mulsanne was designed and engineered in Crewe and – while the debut was orchestrated somewhere that, in contrast, sunshine is virtully guaranteed – that's where it's being made. The bodies and engines come to life there and, overall, the plant is lost in concentration, as Bentley's experts handcraft the interior.

'The new Mulsanne is a thoroughly modern flagship that captures the essence of the Bentley marque,' said Bentley Motors chairman and chief executive Dr Franz-Josef Paefgen. It would be churlish – and wrong – to disagree.

Facts & Figures

'WO' would be proud

The Mulsanne is named after one of the longest straight sections on the Le Mans course. Another rousing chord struck up from Bentley history was the presence of W.O. Bentley's own 8-litre at the Californian launch. Although 79 years older than the Mulsanne, it's still considered by many to be the ultimate incarnation of the marque. This was the second of just 100 made, and also 'WO's company car.

THE BOND BUG

There wasn't much sensible market research underpinning the extraordinary Bond Bug; possibly, you could tell just by looking at its profile, a large wedge of Red Leicester cheese on three wheels.

It was predicated on a realisation, by designer Tom Karen, that no-one marketed a tiny car that was fun to drive and own, but also cheap to run. As he had hard-up students in mind, it would need to be thriftier even than a Mini.

Ogle Design, Mr Karen's consultancy, numbered Reliant – Britain's biggest three-wheeler manufacturer – among its biggest clients, so when he showed bosses there his Bug plans, they could see exactly where he was coming from. They heartily approved when he stressed how easy it would be to build, by using a plastic body on the existing Reliant Regal frame.

The final production car was faithful to Karen's proposal, including the vertically truncated tail, the exposed rear axle, and even a plywood boot lid. If anything, Reliant made the car more radical still by incorporating a lift-up canopy with plastic side-screens, instead of doors and a fixed roof.

Bond Bug publicity stressed youth, novelty, and orangeness!

BOND
THREE-WHEELERS

Bond Minicars, pride of Preston in Lancashire, were a peculiar type of three-wheel, motorbike-engined economy machines launched by Lawrence Bond in 1948. One step up from a motorbike and sidecar, they were initially so basic that the air in the rear tyres doubled for the suspension system. But owners relished their astonishing manoeuvrability and fuel economy that could be eked out to over 100mpg. Reliant, which bought out Bond in 1970, was its nearest rival, although its three-wheelers were always more upmarket, believe it or not.

▲ *Bond Minicar Mk E.*

At £548, the Bond Bug undercut the Mini and Hillman Imp, but not the four-seater Fiat 500 at £498. However, extending Mr Karen's vision of fostering sales among young drivers, Reliant's special 'credit package' offered hire purchase, a two-year warranty and cheap insurance.

Facts & Figures

Two cult 1970s vehicles

The Bug was first conceived by Tom Karen in 1958, when he christened it the Rascal. He had to wait 10 years for the chance to dust off the plans and turn his three-wheeled fun car into reality. By coincidence, in the same year the Bug was unveiled – 1970 – the Raleigh Chopper bicycle went on sale in the UK, another transport icon for youngsters in which Karen and Ogle Design played a big hand.

It was huge fun to drive, nippy, and, once it had taken over 23 seconds to reach 60mph, you could wind it up to 77mph. Motorbike rates for three-wheelers slashed annual tax costs. Plus, of course, it stood out a mile in the only colour scheme on offer – bright orange, with black graphics to emulate racing car sponsorship! 'The sight of this little projectile beetling past more staid saloons produces many a startled look,' reported one contemporary magazine.

Yet Reliant simply couldn't find enough people to take the plunge and buy one. In four years only 2,268 were sold. It seemed any young driver who could afford to run a car wanted four wheels, even if second-hand ones. For such an out-of-the-box British idea, that was a shame.

THE BRISTOL 401

Aeroplane manufacturers never go in for penny-pinching because patchy quality can, literally, be a matter of life and death. That does tend to make aircraft a tad expensive, mind.

This mindset persisted when the Bristol Aeroplane Company decided to enter the car business, with assets adroitly commandeered from BMW as war spoils in 1945. It simply couldn't bring itself to cut corners.

So the £3,112 price of the 1948 401, swelled by Purchase Tax, was about three times that asked for an equivalent Jaguar, and even then it barely covered the cost of making it. Still, it was a fantastic bit of kit: technically advanced, brimming with clever touches, and beautifully made.

Facts & Figures

Advanced details

There was plenty to admire on a 401, yet there was a wealth of unseen detail that proved ahead of its time. This included remote buttons in the cabin to open the boot and fuel flap, an ergonomically shaped steering wheel, and bumpers designed to absorb energy in a knock. The bonnet could be opened to the left or right, or removed entirely, and the door releases were push-button inside and out.

Bristol nicknamed the 401 the 'Aerodyne' after careful attention to streamlining.

After the very BMW-like 400, the 401 relied on an Italian-patented construction method called 'Superleggera' (super-light), featuring aluminium wrapped around an intricate skeleton of curved, small-diameter tubes. There were different thicknesses of metal depending on the position and function of the panel, meaning that a huge amount of handworking was called for. If that was one example of Bristol's high-minded extravagance, then the aerodynamic perfection of the teardrop shape was another permeating its design. Bristol's aircraft aerodynamicists spent ages tweaking the Italian styling in the company's wind tunnel, and in runway tests, until they were completely satisfied. Little wonder they entitled it 'The Aerodyne'.

High-speed silence was a dividend of this attention to streamlining, not to mention up to 25mpg fuel economy. The throaty, straight-six, 85bhp BMW engine could haul the full four-seater Bristol to within a whisker of 100mph, remarkable at a time when a Bentley Mk VI or a Jaguar Mk V needed an engine with twice the capacity to manage that. And the 401 was close to Bentley standards of opulence inside, too, with generous, leather-upholstered seats, and a lovely wood dashboard.

This car was never about all-out speed, though. Fine handling, especially in those pre-motorway days, was even more important, aided by light, precise steering and a wonderfully slick four-speed gearbox. This general feeling of poised good manners won the marque lifelong devotees.

The original Bristol 400 of 1946, owed its origins to BMW, and Second World War booty.

SIX-CYLINDER BRISTOLS

The 401 became the 403 in 1953 with a bit more power, better brakes, and a front anti-roll bar to enhance handling. There was a 402 too, an elegant convertible which, inexplicably, nobody seemed to want, as only 23 were sold against the 401's 650. Bristol built its final six-cylinder car, the 406, in 1961.

THE BRISTOL 411

The V8-engined Bristol sports saloon has an aura of immortality about it. The 407 introduced in 1961 started a line still going strong today, as the current Blenheim 3 is essentially the same car with countless updates gathered along the way.

They're not going to thank you at the company's famous Kensington showroom for pointing it out, but the Blenheim's separate box section chassis, Chrysler V8 engine, four sumptuous armchairs in a spacious cabin, and enormous boot were the exact same hardpoints on the 407 you'd have surveyed there in the early 1960s.

And what's wrong with that? You can call these cars anachronistic but they have a special, distilled character that – while decidedly low-tech compared to a modern Merc or BMW – is still hugely appealing. Every Bristol, engine aside, is hand-built the slow, old-fashioned way in a tiny

Facts & Figures: Mysterious methods

Bristol's survival is down to one man, ex-racing driver Tony Crook, who owned the company from 1960 until 2001 and was still involved – aged 87 – until 2007. He was famously secretive and deliciously crotchety to outsiders. In a memorable interview with Neil Lyndon in the *Sunday Times* in 1988, Crook declared, 'I'm a great anti-overhead man, myself.' He called his clientele 'our people' but, while being interviewed, ignored a potential customer banging on the showroom door . . . and then refused to answer as the exasperated man phoned for attention on his mobile.

The 411 is the epitome of Bristol character, power and restrained style.

Bristol (city) factory. When you buy one you enter the micro-climate of the Bristol 'way' of doing things. It's a society most of us are too cash-strapped to qualify for.

Sorry: a third of the way through and the 411 hasn't even been mentioned. In the context of all modern Bristols, it's the finest of the lot. It's got the meatiest engine, the powerful 6.3-litre (later 6.6) V8 replacing the earlier 5.2, with a limited-slip differential to control the 30 per cent power boost. And, from the Series 3 onwards, it has the best styling, with four intimidating 7in diamater headlights and a grille like a barbecue. Subsequent Bristol saloons can't match its starchy style.

It's a 143mph car, so something of a bruiser in a Savile Row suit. You could drive a 411 recklessly, but it would be improper. The suspension can be individually tuned, the track is narrow, betraying the age of the chassis in a good light, the power steering is light but accurate, and it's an uplifting way to travel fast on sweeping and twisty bends. For such a big car, the 411 has terrific roadholding.

➤ Today's Blenheim, right above, is descended directly from the 407 of 1961, right.

THE GREEN BRISTOL

Bristols, for their size and weight, offer pretty decent fuel consumption, but they could never, ever be called economical. Nonetheless, the British manufacturer has one pioneering eco-credential. It was the first to offer a factory-fitted conversion so its cars could run on liquid-petroleum gas, or LPG, to cut noxious emissions and fuel costs. That was in 1977, and the Blenheim 3 can still be ordered so-equipped.

THE BRISTOL FIGHTER

Bristol Cars is resolutely *Not Like Other Carmakers*. Little surprise, then, that when it decided to create a supercar – in fact, its first totally new car since the 1940s – it would do so in an astonishingly individual manner.

Not for Bristol the ridiculous notion of putting the engine in the middle, racing car-style, and asking the driver to recline as if on a St Tropez sunlounger. The Chrysler V10 engine is upfront in a separate chassis, with a massive steel safety cage around the passenger compartment.

The hand-formed aluminium body panels clad a steel inner structure, all except for the rakish 'gullwing' doors and the tailgate, which are made from stiff, lightweight carbon fibre. Inside, the luxurious cockpit can accommodate the tallest drivers easily, the driving position electrically adjustable to suit, who are snug in a padded cocoon of hand-stitched leather and surveying a bank of instruments that wouldn't look out of place in a Cessna – one of them informs you how many engine hours you've used!

THE MOST POWERFUL PRODUCTION CAR ON EARTH

The Fighter was launched with quiet confidence in 2004. Two years later, after apparently being bombarded with demands from customers for more power, Bristol came up with the Fighter T, nonchalantly mentioning that its twin-turbocharged 8-litre V10 engine produced 1,012bhp. This made it the world's most powerful standard production car, the first with a turbo V10, and trouncing the Bugatti Veyron. But with typical British reserve, the 270mph potential top speed was electronically limited to a 'more than adequate' 225. How sensible. The car is, says Bristol boss Toby Silverton with meek understatement, 'incredibly good fun'.

The Fighter side by side with the 400, the car that began a great British marque.

Your daily transport?

Bristol is at pains to explain what a practical everyday car the Fighter is. Each one has a suspension package tailored to the needs of the owner, you can choose the size of fuel tank you want, and there's a proper, full-size spare wheel yet still, also, lots of luggage space. Bristol has even engineered in a tight turning circle for the rack-and-pinion steering, and can vary the level of power-assistance to just how you like it.

▲▲*Gullwing doors lead on to a leather-lined interior like no other.*

▲ *Headlights do without fairings so the beams can perform better.*

As Bristol called in ex-Brabham designer Max Boxstrom to shape the car, it's no surprise it has phenomenal aerodynamics, with a drag factor of a mere 0.28. With plenty of scientific calculations, a teardrop form and a low centre of gravity, it's a super-slippery machine which still somehow gets away with decent ground clearance of 8in and does without any ungainly aerofoils to pin it to the tarmac. Ever practical, however, Bristol refused to fit faired-in headlamps because they harmed beam accuracy, even if they would have added 1.5mph to the top speed.

Ah yes, the top speed. In standard form it's a 210mph machine, with up to 550bhp on tap and acceleration from 0–60mph in 4 seconds . . . which can be reached in first gear. Yet, with deep reserves of torque to exploit, it's cruising at 100mph at a positively leisurely 2,450rpm.

THE CATERHAM SEVEN

Lotus founder Colin Chapman astutely realised sports car fanatics yearned for a sophisticated but cheap-to-run dual-role car; something to drive to a race meeting, campaign with vigour, and then head home in afterwards. That's why he created the Mark Seven (see panel). By 1969 it had evolved into Series III form, but Chapman was anxious by this time to escape Lotus's humble, kit car origins. So in 1973 he gladly flogged the manufacturing rights to Caterham Car Sales.

Chapman insisted the new owner adopt a different name for the car, but company founder Graham Nearn noticed customers referring to it simply as the 'Caterham Seven'. The title stuck. He later admitted he wouldn't have readily selected the name of a

Not hard to see why fans call the Seven a four-wheeled motorbike.

Facts & Figures

The Lotus Seven

First came the Mk VI in 1952: as Lotus's first real production car, it offered a spaceframe chassis, skimpy alloy body and coil-spring suspension. It was supplied as a kit for home assembly, leaving the constructor to source a second-hand Ford engine and other components. The Seven that replaced it five years later mixed the best of the Mk IV and Lotus Eleven. The basic kit cost £536. Handling and roadholding, benefiting from a low centre of gravity, new tubular spaceframe chassis, sophisticated wishbone suspension and minimal plastic body, were little short of incredible. A vast array of engines could be installed: with a hot Coventry Climax unit, the car was christened 'Super Seven'.

SIXES AND SEVENS

The Seven's most vivid appearance (in its original Lotus guise) in British cultural history was as the car driven by renegade secret agent 'Number Six' in TV's mystery-adventure series *The Prisoner*, broadcast in 1967/8. One of the cars used for filming was supplied by Caterham, and company owner Graham Nearn makes an uncredited appearance in one episode as a humble mechanic.

Patrick 'Prisoner' McGoohan and Graham Nearn.

It's a thrilling track day car.

Surrey commuter town. Caterham has continued to sell the Series III as the Caterham Seven ever since. In fact, the older it's become, the more popular it's been.

Choosing your preferred power unit is a part of the Caterham experience. Over the years, the menu has included Ford Kent, Lotus twin-cam, Vauxhall 2-litre 16-valve, MG/Rover K-Series and Honda's Blackbird motorbike engine. The most extreme of the traditional Sevens was the Superlight R500, with 230bhp from an MG X-Power 1.8-litre K-Series. At 3.4 seconds for the 0–60mph sprint, it was the world's fastest-accelerating car, with male pattern baldness triggered as your hair was wind-whipped at its 150mph top speed. In 2005, the comprehensively updated Seven CSR offered 2.3-litre Ford Duratec power, custom-uprated by Cosworth for up to 260bhp. Among other improvements was the Seven's first ever ergonomically shaped dashboard.

The Nearn family owned Caterham Cars until 2005 when, ironically, it was bought by a consortium of former Lotus managers. It's still headquartered in Caterham, although the cars are now made in Dartford, Kent. It received the Queen's Award for Export in 1993 for helping to buoy Britain's economy.

Lotus never reckoned the Seven would be on enthusiasts' wishlists four decades later. But it remains the ultimate weekend thrill machine; the critic who declared it a 'four-wheeled motorbike' for its phenomenal cornering agility was spot-on.

THE CHRYSLER/TALBOT ALPINE

Hey, just a second! Don't turn the page in disgust. The Alpine may not be much cop to look at now; the unrefined racket from under the bonnet was never a good calling card, either. But it was, for its time, an adventurous family car, and a revolutionary European collaboration in which Britain played a major part.

This handsome, five-door hatchback was a genuine Anglo-French co-production. America's Chrysler imposed an *entente cordiale* on its two European outposts, Simca in Paris and the former Rootes Group in the West Midlands, and the resulting Alpine of 1975 was well resolved. The French side provided the dependable, if coarse, four-cylinder engines and four-speed gearboxes. Meanwhile, *Les Rosbif* in Coventry

Facts & Figures: The Horizon

The Brits and the French reconvened in the mid-1970s to concoct a credible rival to the Volkswagen Golf, and this was it – the Chrysler Horizon. Like the Alpine, the neat five-door design came from the UK and the engineering from France. It was revealed simultaneously in Paris and Detroit on 7 December 1977, for an outwardly similar (but, internally, almost entirely different) Plymouth Horizon was built in America. On sale in Britain in October 1978, it somehow managed to repeat the Alpine's feat of being voted European Car of the Year. Again, like the Alpine, no-one seemed to care.

The hatchback Alpine won the Car of the Year trophy.

The first Horizon is built in Coventry.

carried off an excellent job with the state-of-the-art interior and exterior styling.

The Frenchies refused to sell Chryslers, instead naming the car the Simca 1307, while for the British market the sporty Alpine name was dusted off and reused. Also, as manufacture got underway in Paris, Chrysler UK was begging for a cash lifeline from the UK Government, so Alpines didn't start rolling out of Coventry until mid-1976.

The Alpine previewed family cars of the future. Its front-wheel drive gave surefooted handling and a spacious cabin, while the rear tailgate and folding back seats were very practical. The old-fashioned market leaders, the Ford Cortina and Morris Marina, offered nothing remotely close, and it was crowned European Car of the Year in 1976. But British buyers remained stuck-in-the-mud, preferring the convention of four-door saloons and a separate boot. Against the mighty Cortina, too, the Chrysler took a kicking, its top-range 1,442cc engine outshone by Ford's gutsier 1.6- and 2-litre motors.

All European Chryslers were renamed Talbots by 1980, after Peugeot acquired Chrysler's teetering European outfit (nominal price: $1). The Alpine continued to leave Cortina Man indifferent. But it was an important bridgehead between old and new technology in mass-market cars, and Britain played its part in that.

IDEAL FOR DRIVING DOWNTOWN

The other important British Chrysler was the Sunbeam. Launched in 1977, it was a cleverly shortened Hillman Avenger with a glass tailgate and a new front. Development took a lightning 19 months, as a taxpayer-funded way to keep Chrysler's Scottish plant going. It was surprisingly successful. Petula Clark advertised it on ITV, imploring us to 'Put a Chrysler Sunbeam in Your Life'. She later turned the advertising jingle into a minor hit, rewritten as 'Put A Little Sunshine In Your Life'.

THE DAIMLER 4hp

The Daimler Motor Company was the very first established in this country specifically to make cars. Incorporated on 14 January 1896 in Coventry, the plan was to assemble rear-engined, belt-driven Daimlers from kits of parts imported from the famous German automotive pioneer Gottlieb Daimler. Ultimately, though, the only thing the fledgling business brought over from Bad Cannstatt was the name, and even then the pronunciation was changed from 'Dime-ler' to 'Dame-ler'.

For Gottlieb Daimler turned against supplying the new venture, and the British Daimler company instead copied a much more advanced French car, the Panhard & Levassor. This featured a front-mounted,

One hundred years separate these two Daimlers, photographed in 1997.

Facts & Figures: Royal seal of approval

The 4hp was quickly joined by a more powerful 6hp model. One was delivered to the father of the present Lord Montagu of Beaulieu in 1898. Its steering was by tiller, drive by chain, and a four-speed transmission was operated by two levers on dials, mounted on a metal pillar beside the driver's seat. In January 1900, the Prince of Wales took delivery of a Daimler 6hp Mail Phaeton. This was the ultimate celebrity endorsement, and became the first of dozens of Daimlers purchased by the royal family throughout the twentieth century.

twin-cylinder engine, thoroughly redesigned by Daimler's chief engineer James Critchley, and sliding-gear transmission, which became the template for drivetrain layout for many years afterwards. In November 1896, an artist's impression of 'the first British-built Daimler' appeared in the press. No-one knows if it actually functioned but, if so, it would have been the first car ever with a bevel-drive rear axle instead of belts or chains.

The first all-Coventry Daimler, a 4hp machine, took its maiden run on 2 March 1897, driven by works manager A.J. Drake. Three months later, three were apparently being built each week. The price was £350. This led to Britain's first road test write-up, after two female journalists from *The Gentlewoman* drove one from Northampton to London, noting 'the colossal appetites' that motoring gave them, probably after winding the Daimler up to its top speed of 12mph. A few weeks later, Daimler director Henry Sturmey (he was also editor of *The Autocar* magazine, a flagrant conflict of interests) undertook the first ever car journey from John o' Groats to Land's End aboard his 4hp. He wrote a glowing report of the 1,600, trouble-free miles. . .

➤ *Daimler kicks off car advertising in Britain.*

➤➤ *Inside Daimler's 'Motor Mills' factory in Coventry.*

DAIMLER'S SHADY HORSE-TRADER

The Daimler Motor Company grew out of the Daimler Motor Syndicate founded, in 1893, by the visionary Frederick Simms to exploit Germany's novel internal combustion technology. This business was sold in 1895 to dodgy company promoter Harry Lawson, as the nucleus of his masterplan to control the emerging British motor industry. Lawson, however, soon got distracted by other schemes, including the grandiose Great Horseless Carriage Co. and 'investment opportunities' from American huckster E.J. Pennington. He resigned from Daimler in October 1897, and narrowly managed to avoid losing his shirt in the collapse of Coventry's bicycle industry shortly afterwards.

THE DAIMLER DS420 LIMOUSINE

The Daimler Limousine has been a familiar pillar of British life for donkey's years. We often encountered these black-painted, seven-seater giants on sombre occasions. Many were bought by undertakers as mourners' cars, while a fifth of all DS420s were adapted into hearses. Alternatively, painted cream or white, these Daimlers enhanced the magic of the better class of wedding. If you're giving your daughter away, it's a nobly fitting way to arrive at the handover.

Local councils bought them (at taxpayers' expense) for mayors to swan around in. Embassies in London ran them too. They were adored by foreign royalty. And Hong Kong's Regent Hotel (now renamed The Intercontinental) once operated a 22-strong fleet.

One of the last DS420s, for years a dignified presence at British weddings and funerals.

The Queen and Queen Mother both owned DS420s, reviving the royal patronage Daimler ceded to Rolls-Royce in the 1950s. The late Queen Mum, Gawd bless 'er, had five in succession, while the Royal Mews has three on active stand-by today.

The DS420 dates from 1968. It utilised the Jaguar 420G floorpan with wheelbase stretched by 20 inches.

Facts & Figures

The Daimler Majestic Major

The Majestic Major offered shattering performance for something that looked like a motorised hunting lodge. Under that haughty bonnet was an all-aluminium, hemi-head 4.7-litre V8 – more American hot-rod than British establishment. There was 120mph potential, and 0–60mph acceleration in just 9.7 seconds. What's more, roadholding was none-too-shabby, and with disc brakes and power steering this underrated, upright saloon car could be driven with vigour. The DR450 limousine edition, the DS420's predecessor, added a stretched wheelbase and an extra row of seats.

➤ *The big Daimler's other life was as capital VIP transport for diplomats and royalty.*

It kept the Jaguar 4.2-litre straight-six until 1992, making it the last production car using the Jaguar XK engine that had propelled the D-type to Le Mans victory. Would that fact ever have impressed the bereaved as they passed slowly through the cemetery gates? Probably not.

FIT FOR A PRINCESS

This Daimler DE27 was given to Princess Elizabeth as a wedding present by the RAF and WAAF in 1948, paid for by voluntary donations – 2*s* 6*d* each from officers, and sixpence apiece from individual airmen and women. It was great for Daimler: as Queen Elizabeth II, she became the very best kind of regular customer.

Chauffeurs spread its nickname of 'The Old Lady'. This was about as affectionate as it got, for while the passengers could sprawl in opulent comfort on the 6ft-wide rear seat, the driver was squashed in behind his glass division on a cramped bench seat. Until 1984, this had no adjustment features at all, keeping the employee in his uncomfortable place.

Always handmade and immaculately finished, for many years the DS420 retained outmoded features from 1960s Jaguars, like a column gear selector and a thin black plastic steering wheel. The all-independent suspension gave a serene ride, there were disc brakes, and when required – a dash to Heathrow to catch Concorde, maybe, or whisking VIPs away from paparazzi – decent performance, culminating in a 110mph top speed. Tight bends, though, were best avoided.

There was only one traditional, purpose-built limo alternative to the DS420: the Rolls-Royce Phantom VI. But at £350,000 in 1989, the £30,000 Daimler was incredible value.

THE FORD MODEL Y

If you've ever wondered how Ford got to dominate Britain's car market then meet the car that started it all: the Model Y.

That's right: Henry Ford's Model T is the subject of all the legends and the worldwide acclaim. But from 1920, its British popularity took a pounding. The Draconian rules of the new Motor Car Act levied an annual tax of £1 on each RAC-rated horsepower (as opposed to brake-horsepower). Model Ts had large, unstressed engines for covering vast American-style distances, and were rated by the RAC at 22hp. So the Model T, utilitarian and cheap to buy, suddenly became costly to run. For the subsequent Model A, Ford came up with a rule-bending small-bore engine especially for Britain, cutting the tax from £22 a year to £15. But when the Depression hit in 1929, sales of this thirsty and gutless Model A imploded anyway.

Henry Ford had had 15 popular British small cars shipped to the USA in 1928 for his close examination, but he was reluctant to build a car specifically to suit the United Kingdom. He finally relented but it wasn't until October 1931 that his Detroit staff began work on a proper rival, with the old boy himself helping to build the prototypes. Amazingly, just five months later, the Model Y was unveiled in London.

➤ The stylish Model Y, conceived in a mere five months, handed Ford the key to unlock the British car market.

Facts & Figures

Popular, and long-lived

The little Y-type Ford brought frugal motoring to many Brits for the first time. After a cosmetic revamp in 1937, the car was on the market as either an Anglia or a Popular right up until 1959, by which time it was little short of a living antique. As the world moved on, these cars became glaringly uncomfortable and slow, and the final Popular, despite getting a bigger 1,172cc engine in 1953, got by with a single windscreen wiper and mean little headlights, while the lack of indicators made hand signals a constant chore. . .

There was an all-new, and resolutely unadventurous, four-cylinder engine of 933cc RAC-rated at 8hp, an all-new transverse-sprung chassis, and an elegant, practical four-seater saloon body. It looked good and drove well, thanks partly to being the first sub-1-litre car with synchromesh gears.

By August 1932, Model Ys were flowing down the production lines at Ford's newly erected Dagenham factory. The £120 Model Y gave the Ford Motor Co. instant dominance in the crucial 8hp market sector, and alone saved the Essex-based enterprise from closure. It truly set the company on its unstoppable path to British market leadership.

THE £100 CAR

When desperate rivals Morris and Singer copied the Model Y's natty styling in 1935 to keep up, Ford slashed its profit margins to the bone to create Britain's first and, so far, only £100 saloon car; you can see why they decided to call *that* one the Popular. The only other £100 car was the Morris Minor in 1931 at the height of the Depression, a stripped-out open two-seater.

THE FORD ZODIAC MK II

Smooth, throaty straight-six engine, two-tone paintwork and upholstery, three-speed column gearchange so you can get three people on the springy front bench seat, whitewall tyres, front spotlights, lashings of chrome, sleek yet restrained Americanised lines; for a huge swathe of middle-class Britain, the Mk II Zodiac represented motoring power and luxury within affordable reach, with a starting price in 1956 of £969. Furthermore, for £98 less you could get a Zephyr, with all the same basic attributes like a 90mph top speed, six-seater capacity and an optional overdrive or automatic transmission – Ford's first.

The Autocar magazine was dead right when it declared: 'It is one of the best and most encouraging British cars in full-scale production.' It was alluding to the fact that the big Fords offered some quite advanced features, like MacPherson strut independent front suspension and rugged, well-designed monocoque construction. Direct competitors from BMC, Vauxhall and Standard couldn't hold a candle to Dagenham's finest, while some more costly Daimlers and Humbers were nothing like as agile and satisfying to drive.

The Mk II Zodiac offered advanced features and undeniable style that put rivals in the shade.

You could also buy a Zodiac as a two-door, five-seater convertible complete with power top, or a five-door estate car. Zephyr estates, in particular, became a fixture of long-distance driving up north after the Lancashire Constabulary introduced them to patrol the Preston Bypass, Britain's groundbreaking first stretch of motorway, in 1958. They were the first cop cars painted white, and the livery was quickly adopted by the five police forces covering the M1.

For those on a much tighter budget, there was also the Consul Mk II, largely the same vehicle with a four-cylinder 1.7-litre motor. It was plainer, noisier and slower, but still massively popular.

Having set out ahead of the pack, Ford never got complacent, and made sure the Mk II range kept improving. The most radical change was to trim the car's height in 1959, squashing the roof profile for the new Lowline series that made the Zodiac even sharper. There was the worthwhile option of front disc brakes in 1960, too.

The Mk I cars

The 1950 Consul was the first really modern post-war British saloon, ushering in a whole raft of new technology. Out went a separate chassis frame, and in came unitary construction for the first time, making for a lighter, stiffer structure. Front suspension was by MacPherson struts – a world first for the independent system perfected by American Ford engineer Earle MacPherson – and there was a new 1.5-litre, overhead-valve engine. The slab-sided styling was nevertheless a step-change away from the 1940s norm, with separate mudguards and headlights now eradicated. The six-cylinder Zephyr arrived a year afterwards, with its 2.2-litre straight-six, and the first luxury Zodiac in 1954.

WIDE OF THE MARK

The later Mk III and IV Zephyr/Zodiac ranges were enlarged to become progressively lower, longer and wider cars, shifted upmarket to allow the insertion of the Cortina and Corsair into Ford's range. They may have been dynamically better but, compared to their contemporaries, they somehow lost their attractive edge – something not regained until the German-designed Granada came along in 1972.

THE FORD ANGLIA 105E

Throughout the 1950s, Ford's Anglia 100E had provided cheap, thrifty transport to hundreds of thousands of the nation's motorists. In all likelihood, it was Britain's best-selling car, although in those days nobody was organised enough to laboriously collate monthly and yearly data, so it's impossible to be certain. A sparkling performer, though, it certainly wasn't, with a gutless sidevalve motor rowed along by a three-speed gearbox.

The Anglia 105E, its 1959 replacement, was a gigantic leap forward – a startlingly different small car. For instance, no-one could ignore that angular 'Breezeway' rear window treatment. There was a touch of pseudo-science behind it; it was supposed to keep the rear screen clean and offer greater headroom. Mostly, though, it simply made the crisp little Anglia stand out from the dumpy crowd. Meanwhile the broad front grille and hooded headlamps gave the Anglia a characterful 'face' . . . one now familiar to a whole new generation after the car's appearance in the *Harry Potter* books and films.

Anyway, compared to the opposition, the Anglia was some hotshot: a 76mph top speed and 0–60mph in 16.5 seconds was exceptionally fast for a little saloon. The new-found liveliness was all thanks to an inspiring new overhead-valve engine, one that redefined what

➤ *The Anglia, crisp as a new fiver, had a brilliant engine that just begged to be revved!*

The Kent engine

The first of the long-running 'Kent' series engines arrived with the Anglia. Although compact, it had very 'oversquare' dimensions (the bore was almost double the stroke). This allowed tremendous potential for expanding and tuning, and made for excellent robustness. It was soon a favourite for Formula Junior race preparers, while the Anglia itself, despite being mostly outclassed by Minis in European rallies, came second in the 1963 East African Safari Rally. Ford made over 10 million Kents (and plenty more of the bigger, V6 'Essex' power units). It proved so durable that descendants powered the Ford Ka right up to 2008.

THE FIRST SCOUSE MOTOR

It cost Ford £70m to build a new factory at Halewood, east of Liverpool, forced into the 329-acre site under Tory government pressure to reinvigorate unemployment blackspots. Still, Halewood had good rail links to Dagenham (for engines) and South Wales (axles). On 8 March 1963, Liverpool's Lord Mayor drove the first car, an Anglia 105E Deluxe, off the production line. Sound.

drivers could expect from a cheap (£589 in 1959) small car. This one seemed to actively enjoy being revved!

Another first was a four-speed gearbox – amazingly, all previous Ford road cars had had just three forward speeds. Ford offered an enlarged 123E version of the engine in the Anglia Super from 1962, which gave 13bhp more than the standard Anglia's 53bhp. The Super spec also included chrome flashes, carpets and a heater. There was one other bodystyle, an attractive estate, from 1961.

The Anglia succeeded in becoming Ford's first million-seller in Britain. It was easy to see why. It was more comfortable than a Mini, more reliable than a Hillman Imp, and tougher than a Triumph Herald.

THE FORD LOTUS CORTINA MK I

Car enthusiasts were gobsmacked at the boldness of this idea – a family saloon given the 'full works' treatment by Lotus. And not just any old saloon, either, but the compelling new Cortina at that. Blimey!

Ford's boffins made the 1962 Cortina as light and spacious as possible, designing it to use the minimum of metal to make the car best suited to Britain's new, press-on motorway age. But even with the biggest 1.5-litre engine, it was pleading for more oomph. And that's precisely what Lotus provided. The future F1 champions installed a trick new engine, a high-performance twin-cam that initially used basic Ford components but quickly evolved into Lotus's own bespoke confection. With 1,558cc, twin carburettors and 105bhp on tap, it would have transformed any ordinary Cortina. But Lotus wasn't finished.

Steel doors, bonnet and boot lid were replaced by aluminium items, there were power front disc brakes, and all-new coil-spring rear suspension to keep the Cortina's jumpy rear end tethered to the tarmac. Outside, it was easily identifiable with its cut-down front bumpers, wide wheels, and white paintjob enlivened with green flashes.

Facts & Figures: Britain's best-selling Cortinas

Between 1962 and 1982, 4,279,079 Cortinas were sold in five distinct series. Each type sold over a million, and these ever-dependable, if trenchantly conventional, cars officially topped the British sales charts in 1967, 1972–5 inclusive, and 1977–81 inclusive. The Cortina was in second place for eight years and in third for the remaining two.

➤ *Wide wheels, tiny bumpers and a green flash broadcast the Lotus energy inside this two-door Cortina.*

Okay, it cost £1,100, double the price of the most basic two-door Cortina, but it could reach 108mph and its near-perfect balance and independent rear suspension meant it could positively tear round corners.

The Lotus Cortina set race tracks alight. In 1964, F1 World Champion Jim Clark took the British Saloon Car Championship in one, while Sir John Whitmore won five European fixtures. The so-called 'Bearded Baronet' was synonymous with the Lotus Cortina's on-track antic of cornering with one front wheel off the ground. In 1965, Whitmore completely dominated the European Saloon Car series and in '66 he swiped four more outright victories in Europe, while Clark won three in Britain.

The Lotus Cortina would always be rare, with just 3,301 sold until 1966 – a tiny fraction of the Mk Is. And in 1965, the fragile coil-spring suspension was downgraded to cheaper, more robust Cortina GT-style leaf springs. But it was only the beginning of Ford's sports saloon heritage.

STYLE ICON OF THE 1970s

Once, no-one even noticed the 1970–6 Mk III Cortina, as it was such a mundane, everyday sight. And it was a fairly unremarkable machine, popular for a host of reasons that included things like value, roominess and a vast choice of options: you could have a two- or four-door saloon, an estate, engines from 1.3- to 2-litres and a brain-frying roster of trim levels. Now that almost all of them have been cubed and smelted, though, the Mk III – with its 'Coke bottle'-shaped wavy styling, and associations with iconic TV cop shows – is being admired anew.

THE FORD ESCORT

European Fords have always epitomised simple, reliable, cheap-to-run and easy-to-maintain motors. They've tended to offer the perfect fit for the typical driver, and that's something that British thinking helped define in the 1960s. The hugely popular Escort is a prime example.

After hitting the bullseye with the 105E Anglia, its replacement – the 1967 Escort – suited economy motorists even better. Like the Anglia, it was very conventional, a rear-drive saloon with a four-cylinder engine and an old-fashioned, cart-sprung live back axle. But it was lower, sleeker, more American-feeling, with rectangular headlights and a front grille shaped like a cartoon dog's bone. Ford declared it 'The small car that isn't', and buyers agreed, piling into Ford dealers to get one.

Moreover, the newly formed Ford of Europe organisation found the Escort – created entirely in Britain – such a solid proposition that it was sold all over the continent. There was clearly a sporty car trying to get out here. You could tell that by the excellent rack-and-pinion steering and predictable handling in even the most basic Escort. Ford was happy to oblige.

It started off with a mild, 75bhp Escort GT, ideal for feisty pensioners, but then offered the 105bhp Twin-Cam, with a Lotus-type engine and tauter suspension, which was duly replaced by the RS1600 with a 120bhp 16-valve Cosworth engine.

Facts & Figures

Here comes Brenda . . .

The Mk II Escort of 1975 was merely a sharper, tidied-up version of the original. They called it 'Brenda' inside Ford while readying it for launch; it was getting a bit long-in-the-tooth but was as solid a machine as ever, and now ran to a Ghia luxury edition too. Here are two British institutions: a Mk II Escort and anarchic parliamentary candidate Screaming Lord Sutch, about to take part in a celebrity saloon car race.

The hot Escorts proved awesome rally cars. The Twin-Cam saw victory in the 1970 London–Mexico Rally, while the RS1600 dominated British rally and saloon car racing throughout the 1970s. Ever quick to capitalise on its motor sport success, Ford rushed 1.6 Mexico and 2-litre RS2000 Escorts into showrooms, so customers could own a piece of the lairy action themselves complete with wide wheels and, naturally, go-faster stripes.

The Mk I Escort was never the most advanced car on the street, but even the four-door saloons and estates had a whiff of high-performance charisma about them.

LEADING FROM THE FRONT

Ford's response to the runaway success of the Volkswagen Golf was to follow suit, putting the Escort name on a front-wheel drive hatchback in 1980. It had a brand new range of CVH engines, built at a new factory at Bridgend in Wales, and a fresh image-leader in the XR3 – soon to be the default 'hot hatchback' choice of the British boy racer. This competent range enjoyed massive 1980s success. From 1990, the next-generation Escort proved another big seller but lacked sparkle, leading Ford to pension-off the Escort name permanently in 1997.

➤ *Hot Mk I Escorts, like this RS1600, got the sporting headlines, helping humbler versions get the customers.*

THE FORD CAPRI

To venerate the Ford Capri can seem a trifle hollow. After all, it was a ruthlessly calculated enterprise, its sporty persona precisely cut to the millimetre, its design details distilled by committees. And yet a car emerged that us Brits, especially, took to our hearts for its easy style, four seats, low running costs and often macho performance. We loved the Capri so much that, in its twilight years of 1985 and '86, it was made solely for the British market.

Here was a sporty – as opposed to 'sports' – car tilted at baby boomers. The engineering was straightforward because the Capri shared many unseen components with the Escort. That meant a front-mounted engine and rear-wheel drive. Provision was made for 2- and

CAPRI II AND CAPRI III

It proved tricky to repeat the impact of the first Capri. When, in 1972, Henry Ford II viewed a prototype for a Capri replacement without the car's distinctively rounded rear quarter-window, he barked: 'This ain't Capri', and demanded a redesign. Hence, the Capri II of 1974 merely updated the basic profile and offered a hatchback, and the 1978 Capri III sharpened the looks up with four headlights. Performance pick of the IIIs was the rapid 2.8 Injection.

➤ *The unmistakeable Mk I Capri with fancy wheels, fake air intakes and famous 'hockey stick' crease.*

3-litre engines never offered in the Escort, but even the lowly versions were splendid fun to drive.

The groovy fastback styling was a winning amalgam of wide-opening doors, low roofline, and a distinctive side profile including that hockey stick-like crease, rounded side window shape, and fake air intakes. The cockpit took plastic and fake wood and melded them into a pseudo-sports car cocoon.

To the fortunate few owning exotic Lancias and Porsches, the Capri was a neo-Detroit horror. To the many who could never afford such cars, however, the trusty Capri was the answer to a prayer. Ford's advertising slogan for the January 1969 launch: 'The Car You Always Promised Yourself', was as sharp as the Capri's black vinyl roof and pretend alloy wheels. In 1970 alone, one in four of all European Fords sold, almost 250,000, was a Capri. The millionth was built in August 1973, although it was only British-made (on Merseyside) until 1976.

The Capri was a true motoring democratiser. Just as the Model T put the common man on wheels, so the Capri gave the hard-up family motorist everyday motoring thrills.

The typical Capri

Retrospectively, much is made by devotees of the 3-litre Capri. Yet the typical Capri sold in Britain was actually a 1600 with L, GL or XL trim and manual gearbox, and this would struggle to nudge 100mph. Most buyers weren't overly concerned about high performance in 1969, although they loved the prices: the base 1300, a car for the truly timid or frugal driver, cost just £890, while even a peppy 2000GT was a mere £1,088. The range was so enormous that no dealer could stock every variation. Front disc brakes and excellent rack-and-pinion steering were common factors, but initially you had to fork out extra for seatbelts.

THE FORD SIERRA

Every now and again, a car makes an impact well beyond its actual significance, and the word 'Sierra' was high in the national consciousness in 1982. British car-buyers were alarmed at how expectations were being overturned by this large family saloon, and that the future was being imposed on them in a manner they found unsettling.

In the eye of the storm was the Sierra's curvaceous, wind-cheating shape, large plastic bumpers and ergonomic interior. Conservative customers, company fleet buyers in particular, were attached to the predictable, square-cut Cortina that had served British motoring needs for 20 years. They were quick to damn the new car for its preponderant acreage of plastic and the 'jelly mould' profile of its aerodynamic body.

Under the skin, though, the car wasn't so very far removed from the Cortina it was usurping. Although there was a new independent suspension system at the back, the driveline and engine choice was Cortina through and through. The Sierra carping changed tack rapidly; now it was deemed behind the times, rather than too far ahead of them.

And yet the Sierra did innovate, as self-levelling suspension featured on the 2.3-litre Ghia estate, and the XR4x4 helped popularise all-wheel drive for conventional road cars. As for sporty versions, the 2.8-litre V6 XR4i, with its fuel-injected 150bhp, stiffer suspension and bi-plane rear spoiler stoked enthusiast interest. Later came the feisty Sierra Cosworth. The Sierra ultimately enjoyed a successful 11-year production run, during which it single-handedly kept Ford's enormous Dagenham factory thriving, with just short of 3.5 million examples sold. No wonder 'Sierra Man', to represent the Middle England swing voter so crucial to general election victory, entered the Westminster lexicon; Tony Blair simply updated it to 'Mondeo Man' to clinch his 1997 win, despite the fact that the Mondeo was never a British-made car.

▲ *The Sierra seemed radical at its 1982 launch, but it soon became a yardstick for swing voters.*

Facts & Figures: The Sierra Cosworth

The Cosworth-engined Sierras, with their turbochargers and, latterly, four-wheel drive, represent but a drop in the Sierra ocean; the overwhelming majority of Sierras were 1.6- or 1.8-litre company hacks. Yet the Cosworth gained a larger-than-life notoriety far removed from the saloon car racing circuits it was intended to dominate. As the late 1980s craze for ram-raiding shops and banks snowballed, the Sierra Cosworth was seen as the prime choice, a bad lad's totem, for window-shattering attacks and a speedy getaway. The limited edition RS500 was the ultimate Cosworth incarnation: huge disc brakes, a massive rear aerofoil, and 224bhp from the 2-litre turbocharged engine. That meant 0–60mph in 6 seconds and a 149mph top whack; no Post Office branch felt safe!

SALUTE FROM THE V&A

The Boilerhouse Project at London's Victoria & Albert Museum staged an exhibition about the development of the Ford Sierra in October 1982. They called it 'The Car Programme'. To get a car into the building, a Sierra had to be craned in from above, dangling 70 metres above ground level before being lowered into place.

THE HEALEY 2.4-LITRE

Donald Healey first hit the national headlines in 1931, when he drove an Invicta to a superb win in the Monte Carlo Rally. Sixteen years later, he was back in the limelight, this time as a carmaker in his own right. In 1947, an early example of the brand new Healey 2.4-litre accelerated off towards the horizon along the Jabbeke Highway, Belgium's early version of a proper motorway. With Donald Healey at the wheel, the streamlined car was wound up to 111mph on the dead-straight road, which made it the fastest four-seater saloon in the world, and Britain's first regular production saloon that could manage 100mph without exploding. Shortly afterwards, the intrepid Mr Healey, a diminutive Cornishman, drove another 2.4 to a class win in the 1947 Alpine Rally.

This was something the country, despite the gloomy post-war atmosphere, could really feel puffed-up about, even if the £2,723 price for the luxurious Elliot saloon made it strictly a rich man's plaything; this cost was inflated by double Purchase Tax because the basic car cost over £1,000 (actually £1,750).

It had a light but rigid new cruciform-braced chassis with expensively made front trailing link suspension (inspired by pre-war race-car designs), and was

◄ *A proud Donald Healey at the wheel of a 2.4 prototype; note the concealed headlights, later axed.*

Facts & Figures

The energetic entrepreneur

The ebullient Donald Healey was relentless in his efforts to get his cars built. When it seemed supplies of the Riley engines he needed might end, he set off in December 1949 to lobby Cadillac for some of its V8s. After being rebuffed, he was making his way home dejectedly back across the Atlantic on the liner *Queen Elizabeth* when he chanced upon George Mason, head of Nash-Kelvinator. A drink in the bar led to dinner together, and eventually to the engine supply deal for the Nash Healey sports car. Funny old world. . .

▲ *Elliot saloon and Westland tourer help celebrate the British motor industry's golden Jubilee in 1946.*

MY COUSIN JAMES

One of the original backers of Healey's ventures was James Watt, a former car salesman who, thanks to his wife's inheritance, was able to make the Healey 2.4-litre a reality. He was also this writer's cousin twice removed, which is just about my only claim to fame in the real world of cars. Among his many adventures in his Healey days was a two-day drive from Warwick to Turin to deliver a Healey to coachbuilder Bertone; the car had no bodywork and he wore a white flying suit and goggles for the 1,000-mile-plus trip.

powered by the excellent Riley 2,443cc four-cylinder engine, good for 104bhp. Initially there were two models: the open Westland and the closed Elliot, both with hallmark kite-shaped grilles and bodies built in Birmabright aluminium over an ash frame, although the flip-up headlamp covers on prototypes were dropped.

Healey himself and other contemporary drivers campaigned the cars with great panache in events like Italy's Mille Miglia, in which one won the Touring Car class in 1948. But glory, as is so often the case in the competitive world of 'ultimate' cars, proved short-lived. Jaguar's XK120 comprehensively overshadowed the Healey in 1948 and '49, when it amply proved itself to be a 120mph-plus car. Just under 600 2.4-litres in various forms were sold.

THE HILLMAN MINX

In the car league of its day, the Hillman Minx was never really a pacesetter, and that applied from the time the catchy name appeared in 1932 to the moment when it finally shuffled off this mortal coil in 1970. So why is it in this celebration of British motoring pride, you may pertinently wonder?

Well, year after year, the Rootes Group churned the Minx out of its Coventry factory in tens of thousands, and the successive generations

THE SUPER MINX

The Minx was so popular that replacing it was a thorny problem. The bigger, more spacious Super Minx of 1961 was one idea, but in the end it simply joined the range as an additional model. And for many customers with expanding families or a bit more disposable income, it meant they could graduate to a more grown-up Hillman rather than desert the marque and head for the nearest Ford showroom.

of compact saloons found ready buyers for their combination of reliability, fashionable design and value. They trailed prevailing trends but offered predictable, workaday motoring. They were the Korean cars of their era, long before South Korea actually made any cars, and at a time when cheap imports weren't allowed anyway.

Here's the Minx most people recall today; there had been eight 'Phase' models previously when this became the first of five 'Series' cars. The engine had already been modernised with overhead-valve design in 1954, and now the Series I of 1956 ushered in sculpted new styling and a much roomier interior. In fact, Rootes had called in American consultants to help them make the Minx irresistible to buyers, which explains the wraparound rear window and the two-tone paintjobs. In 1959, the Studebaker look extended to subtle rear tailfins as well.

The Rootes people were well ahead of the curve on choice, too. At a time when most rival models were available in one or maybe two guises, you could get a Minx in Special (which wasn't very) and De Luxe form, as an estate, as a convertible, in a more elaborate livery as the Singer Gazelle, and as the two-door Sunbeam Rapier.

This basic range lasted until 1967, with annual tinkering from designers to keep the cars up-to-date. The

The Sunbeam Rapier

The Rapier, which was really a two-door Minx, actually made its 1955 debut before its Hillman sibling. Apart from the stylistic differences, it had a bigger Stromberg carburettor to help the engine breathe better, although that was quickly updated to twin carbs after Rapiers set out to tackle European rallying. A Series I came fifth in the 1956 Monte Carlo Rally, while a Series II actually won the 1958 RAC event. Rapiers tended towards being solid class-winners and top-10 finishers, with plucky showings in the Monte Carlo and Alpine Rallies right into the early 1960s.

Minx family came to be stitched into the fabric of Britain's roadscape, and as second-hand cars permitted many families to move up to something bigger and better for the first time.

◀ The Series I Minx, seen here in rare convertible form, brought American-style glamour to suburban driveways.

THE HILLMAN HUNTER

If the Minxes on the previous pages stirred some memories, then you may also recall the onslaught the Minx faced in 1962 from, on the one hand, the cheap and lively Ford Cortina and, on the other, the high-tech Austin 1100. The Rootes Group needed to fight back, and it made a decent fist of it with the Hillman Hunter of 1966. To get the job done, the company appointed designer Tony Stevens as its first ever 'product planner'. He was charged with bringing out the 'Arrow' range, more familiar now as the Hillman Hunter, in 24 months '. . . design to showroom – aged 30, I had to sign off every engineering decision on the car. It was totally exhausting.'

Apart from the neat new styling, from Stevens' colleague William Towns, the Hunter was a quantum leap over the older Minxes to drive, with disc brakes and strut suspension upfront, and a reasonably poky – 74bhp – five-bearing 1,725cc engine. As ever, Rootes wanted to cater for all comers, so there was a 54bhp 1,498cc Minx edition stripped to the essentials of any frills, and, in contrast, progressively more upmarket Singer and Humber models. Chuck in the estates and the closely related

The estate was but one versatile member of a huge Hunter clan.

The Hunter (this is a 1972 era model) took a swing at the Cortina, for which it was a worthy rival.

Sunbeam Alpine/Rapier coupés and the vast line-up was completed with customary Rootes thoroughness.

Stevens went on to mastermind Rootes' assault on the 1968 London–Sydney Marathon, which a Hunter won. 'It was one of those rare occasions when a competition result has a monetary value,' he recalled. 'Rootes' share price rose when the win was announced.'

The Hunter GLS of 1972–6 was the dark horse of the range. Its 93bhp engine was souped-up by motorsport specialist Holbay with gas-flowed ports and combustion chambers, a high-lift camshaft, superior pistons, a four-branch exhaust manifold, and twin Weber carburettors. The largely unmodified Hunter, despite a tendency to oversteer on wet roads, took the extra power well, offering sparkling acceleration – 0–60mph in 9.7 seconds and a raucous max of 105mph.

Approved of by Teheran

The Hunter stayed on British forecourts, largely unchanged, until 1979. This feat was sustained by an extraordinary agreement with Iran. Kits of parts to assemble complete Hunters in Teheran were first exported in 1967, and the venture proved so successful that the Hunter – renamed the Paykan – became Iran's national car, staple taxi, and pick-up. Rootes' old Coventry factory dispatched tens of thousands of kits until 1985. Then manufacture switched entirely to Iran, where 40 per cent of registered cars were Paykans. The old Hunter's demise was announced in 2005, whereupon frantic customers placed orders for two years' worth of production!

THE HILLMAN AVENGER

This true successor to the original, compact Minx arrived in 1970. Incredibly, it had six personalities over its lifetime. In 1976, all Hillmans, Avenger included, were rechristened Chryslers, and in 1980 this decrepit Escort-basher was turned into a Talbot. It had been sold in the USA as the Plymouth Cricket and in mainland Europe as a Sunbeam Avenger. When UK production stopped in 1981, the car lived on for 11 more years in Argentina as the Dodge/Volkswagen 1500.

THE HINDUSTAN AMBASSADOR

They call it 'The King of Indian Roads' in West Bengal where the Hindustan is manufactured. Top politicians may have deserted it for armoured SUVs but Sonia Gandhi, wife of late Prime Minister Rajiv, wouldn't be seen in anything else, it's claimed.

The Hindustan Ambassador is a long-standing symbol of Indian national pride. It's in use across the enormous country as taxi and official government car, and it's still the ultimate motoring aspiration for the conservative Indian middle classes to one day own one.

But, as you probably know, the Ambassador's roots are as British as shepherd's pie or, er, that Birmingham favourite, the Balti. Hindustan first took out a licence to build the Morris Oxford Series III from kits in 1957. In 1959, after five years, Morris had replaced the car altogether and was delighted to pack all the tooling off to Calcutta. The car has remained, visually, locked in a delightful time-warp ever since.

Mind you, it's a spacious vehicle, and its unbreakable leaf-spring suspension makes the perfect buffer zone between India's appalling road surfaces and the behinds of the country's gentlefolk.

Improvements came slowly. In 1992, Hindustan finally decreed the wheezy old British engine must go, and installed a 1.8-litre Isuzu lump, with a blazing 75bhp, together with a five-speed gearbox; amusingly,

THE MORRIS OXFORD

The Ambassador is the risen ghost of the old Morris Oxford – specifically, the 1956–9 Series III. There was an Oxford in the Morris range for an awfully long time, beginning in 1913 with the 'Bullnose' and ending in 1971 with the last Series IV. Sir Winston Churchill owned one of these final Oxfords.

◄ Once familiar in Britain as a Morris, the Ambassador remains a much-respected icon of Indian motoring.

that made it the fastest Indian-built car available. And as only Indian-built cars *were* available in India, that made it the local Porsche. Also for the first time in 33 years came optional bucket seats instead of a front bench.

Since then, with the local car market gradually admitting Indian-made versions of newer cars from abroad, the Ambassador has been much modified, getting a new frontage for its 2004 Avigo edition (the old-style car's now called the Classic). There's a 2-litre diesel option, too. The Ambassador is living on time that's gone from borrowed to stolen. Soon, surely and inevitably, it will be no more. But until then, it remains a thing of wonder.

It's Calcutta, not Cowley, but everything about the Ambassador is wedged in a timewarp.

Facts & Figures

The Fullbore Mk 10

In 1993 the Hindustan Ambassador returned to its roots. A couple of entrepreneurs started Fullbore Motors in a Fulham backstreet and, because of the car's emissions-compliant Isuzu engine, could import and sell it. Only a handful were shifted. Fullbore was bankrupted by needing to respray bodywork and fit catalytic converters, anti-roll bars, seatbelts and heaters before the cars were marketable, at an unfeasibly high £11,425. Welsh firm Merlin Garages resumed limited imports in 2002. Notting Hill minicab firm Karma Kabs runs them decked out with silk flowers, smoking joss sticks, glass beads and lucky charms, and they're fully booked by local It Girls and trustafarians.

Behind the veteran visage is Isuzu power and a five-speed gearbox.

THE HONDA CIVIC

You can get several varieties of Honda Civic, depending on which territory you inhabit, around the world, but the European three- and five-door models are probably the raciest of all. These aggressively styled cars, with their sharp, triangulated details, are a universe away from the cuddly Civics that once had principal appeal to smug retirees. The Perspex cover over the grille and front lights, the hatchback split by a manly tail spoiler, an engine start button, and the exposed metal-effect fuel filler all owe more to the pit-lane rather than the crawler lane. Honda has even hidden the rear door handles of the five-door to make it more like a coupé.

There are plenty of quiet, thrifty engines on offer, starting with a 1.4-litre i-VTEC petrol and including a 2.2-litre turbodiesel. The real flier, though, is the 201bhp 2-litre Type-R, with a specially-tuned Vehicle Stability Assist system to make cornering safe and exhilarating simultaneously. All the Euro Civics have electric windows, side curtain airbags and anti-lock brakes as standard kit, and a six-speed manual transmission is offered across the range. And who couldn't love the secret compartment under the boot floor that can also be used to deepen the load bay?

There's a lot to like, and also a lot to be proud of. For these cars are built exclusively in Britain, at the productive plant Honda opened on a disused Swindon airfield. The company chose the site in 1985, building its first car there, an Accord, seven years later. Since then, the Japanese commitment to Wiltshire, where its employees are respectfully referred to as 'Associates', has been laudable. Honda began building engines there in

Two top British performers – F1 ace Jenson Button at the wheel of a Civic Type R.

➤ *Racing car-inspired details – and concealed door handles – have transformed the Civic's formerly wrinkly image.*

1989, pressing bodies in 1995, opened a second assembly plant in 2001, started making its own plastic parts in 2004, and began a new diesel engine casting plant in 2009.

Honda is now a bona fide British manufacturer with 70 per cent of its output exported to 60 countries. In light of Ford's exit from British carmaking, Rover's sad demise, and the closure of Vauxhall's Luton car plant, that's nothing short of heart-warming.

Facts & figures: Two million British Hondas

Honda built its two millionth Honda in Britain on 21 July 2008; a silver CR-V, it was. Among the Hondas made at Swindon are Accords, Civics, CR-Vs and, from 2009, the little Jazz. It was a satisfying day, in 2001, when the first Civic Type-Rs were exported for sale in Japan, as was the moment in 2007 when Swindon built its millionth Civic. The 1981 Triumph Acclaim was actually the second-generation Honda Civic/Ballade built under licence by British Leyland, making the Civic the first Japanese-designed car ever built in Britain.

CLEANLINESS IS NEXT TO GODLINESS

The first Civic in 1972 was Honda's first family car with a water-cooled engine . . . much to company founder Soichiro Honda's dismay, as he always championed unpopular air-cooled engines. The Civic went on sale in 1973 in the US; its CVCC engine complied with the 1973 Clean Air Act and was extremely economical in the face of the fuel crisis. Its popularity made the US Honda's biggest market by 1977.

THE HUMBER SUPER SNIPE

From a Super Snipe's capacious back seat, General (later Field Marshal) Sir Bernard Montgomery – simply 'Monty' to the admiring British public – co-ordinated the battle campaigns that helped edge the Second World War towards its conclusion.

Amazingly, his Humber staff car had already serviced his relentlessly gruelling transport needs since October 1942 throughout North Africa, and would continue to serve his successor in Italy, Sir Oliver Leese, during the rest of the war. It truly deserved its 'Old Faithful' nickname.

Meanwhile, as Monty shifted his focus and tactics in August 1944, a second Super Snipe joined him on the Western Front. 'I have travelled many hundreds of miles in my Humber car,' he wrote about it, in August 1943, in a letter to Rootes Group staff in London, 'and will no doubt travel many more.' An understatement: Monty would eventually cover an incredible 60,000 miles in what became known as the 'Victory Car' over the 12 months to the cessation of hostilities in 1945.

The bravery and determination of the land-based fighting forces is almost unimaginable today. But it was the sheer mechanical tenacity of the Humber that enabled it to play its part. The Super Snipe had barely been in production for a year, with 1,500 or so sold, when war was declared. It mixed the chassis and bodywork of

'Monty' in 'Victory', parading through Wakefield. (Photograph courtesy of Coventry Transport Museum)

the 3-litre Snipe with the 4-litre engine from the stately Humber Pullman. This cocktail produced a fast car, capable of 79mph, tilted at the professional classes and government ministries. However, unlike other executive models, this one continued to be produced throughout the war as a staff car for Army, RAF and Royal Navy commanders.

Surprisingly little was needed to beef Super Snipes up for military life. The bodywork was shortened at the back so they could better tackle ramps, stronger springs were used, and chunky low-pressure tyres were fitted that gave that high-pitched moaning sound later synonymous with Land Rovers. Army bigwigs were won over by the Super Snipe. One officer serving in East Africa covered 48,000 miles on terrible roads in his; 'Both my African driver and I became devoted to it,' he wrote.

➤ *A Humber military staff car.*

Facts & Figures: Old Faithful and Victory

Both the two famous Humber Super Snipe staff cars have survived. 'Old Faithful' was presented back to Rootes in 1945 as a 'token of the good service rendered by Humber vehicles to the war effort'. Its guardian today is the National Army Museum. 'Victory', meanwhile, found a permanent peace-time home in the Coventry Transport Museum, where it remains a star exhibit.

▲ *The visually refreshed 1948 Super Snipe, with Sir Alexander Moncrieff Coutanche Kt, one-time Bailiff of Jersey.*

THE SUPER SNIPE ON CIVVY STREET

In 1946, the Super Snipe returned to Humber showrooms unchanged from 1939, and found a steady market with its powerful, 100bhp straight-six – it was tough, rapid and dependable, but thirsty. In 1948 a modernised model arrived, its headlights now integrated into the wings in the contemporary style, and an imposing two-door convertible was also offered. A slightly improved Mk III appeared in 1950 and continued until the old warhorse was pensioned-off in 1952. Nonetheless, the Super Snipe name survived in Humber's range until 1967.

THE JAGUAR XK120

The XK120 was a divine accident. Jaguar started plotting the all-new Mk VII saloon and also its first engine, the XK, during the Second World War. But when the Mk VII project hit delays, the Coventry manufacturer decided to install the XK motor in a sports car – one that could double as a temporary testbed and a valuable publicity exercise for the engine. A tentative plan to make 240 cars – sufficient to recoup development costs – was drawn up.

The styling, however hastily conceived, was beautifully, classically proportioned. Even today it ranks as a truly timeless shape. Duly completed at breakneck pace, the new XK120 became the undisputed star of the 1948 Earl's Court motor show, carrying the amazing bargain price of £998 (or £1,298 with Purchase Tax added). The publicity stunt worked rather too well: Jaguar quickly decided – owing to clamouring potential customers and a sackload of orders – to change tack and make the XK120 a full catalogue model.

The XK120 roadster took over a year to reach proper production. After that, you still hardly ever

WILLIAM LYONS

Lyons conceived, designed and shepherded each model into production, oversaw the manufacturing process, and astutely handled the day-to-day running of Jaguar. He began in 1921, making motorcycle sidecars in Blackpool, soon progressing to car bodies and then, in 1931, his own SS cars. Lyons adopted the Jaguar name in 1935. After the war, he renamed his company Jaguar Cars, and never looked back.

◀ *At Jabbeke in 1949 with 'Soapy' Sutton and the awe-inspiring, 132mph XK120.*

William Lyons' fixedhead coupé remains one of the world's most beautiful cars.

saw one in Britain because almost all were exported to eager American buyers; subsequent models included a graceful fixed-head coupé in 1951, and a drophead coupé in 1953 with a luxurious mohair hood. A Special Equipment package added wire wheels and 20 more bhp.

The Mk VII-derived chassis featured independent front suspension, with semi-elliptic springs and lever-arm dampers at the back. The manual gearbox was a four-speeder, with synchromesh on the upper three. Braking was by fully hydraulic Lockheed 12-inch drums – barely adequate for such a high-performance car.

Sceptics initially doubted the claims of 120mph top speed alluded to in the car's title, and a 0–60mph time of 10 seconds. But Jaguar test driver Ron 'Soapy' Sutton silenced them when journalists witnessed him reaching 126.5mph in a 1949 demonstration, in an XK120 with hood raised. They then saw the steel-necked Sutton reach 132mph with windscreen and hood removed. It was official: the XK120 was the world's fastest production car.

Facts & Figures: The XK engine

This twin-camshaft, straight-six was a masterpiece. Flexible and powerful, its basis was a cast-iron block with a seven-bearing crankshaft. On top was an all-alloy crossflow Weslake cylinder head. Two noise-reducing timing chains and twin 1.75in SU carburettors completed the specification. Three engineers – William Heynes, Walter Hassan and Claude Baily – designed it all during long nights on wartime fire watch at Jaguar's factory. The resulting 3,442cc engine produced 160bhp at 5,200rpm, and looked superb, with polished aluminium cam covers and stove-enamelled exhaust manifold. It powered the XK120, and its XK140 and XK150 descendants, and was manufactured until 1992.

This famous example, with NUB 120 registration, won the Alpine Rally in 1950 and '51.

THE JAGUAR MK II

The question anyone might reasonably pose, while drooling over the lithe, chrome-encrusted and leather-lined Jaguar Mk II, is: 'What about the Mk I?' No-one talks about the Mk I. It's a Queen Mother/Princess Alice situation: one is a much-loved and vivacious British institution, constantly on parade, while the other is locked away, suspected as rather unstable, and unmentionable.

In fact, there never was a Mk I; not officially. It's a retrospective title for the 2.4-litre saloon, introduced in 1955. Roadholding was okay but it was a sluggish performer. So Jaguar created a 3.4-litre edition in 1957, but the balance now tipped from gutlessness to overpowered: the 210bhp provided sensational acceleration responses but a back axle 4in narrower than the front, and useless drum brakes, made high performance a dice with death. Indeed, in January 1959, British World F1 Champion Mike Hawthorn died in the wreckage of his 3.4 after losing control at over 100mph.

Jaguar needed to salvage the situation, and it came up trumps

with the Mk II. The car that emerged from the revamp was an unlikely all-time great.

The fundamental handling problems were fixed with a widened rear track, new back axle, revised suspension and standard disc brakes all round. The company now felt confident enough to offer a 3.8-litre engine too, conjuring up the ultimate 1960s sports saloon, a blistering 125mph performer with grip to match. In 1961, *The Motor* tested a Mk II 3.4 automatic and declared: 'It offers an outstanding combination of speed, refinement and true driving ease. When price also is considered, it is easy to see why Jaguar competition has been driving one make after another out of existence.'

Jaguar gave the Mk II a facelift even Mary Archer wouldn't mind bragging about. A visual wonder was worked by slimming down roof pillars and elongating the side window shape so it tapered to a curvaceous finish just short of an enlarged rear window.

After a shaky start as the 'Mk I', the Mk II became the *compact sports saloon of the early 1960s.*

THE JAGUAR 240 & 340

With the XJ6 on the stocks, Jaguar was keen to bring the Mk II story to a close. But there was life in the ageing pussycat yet. In 1967, the 3.8-litre engine was dropped and the remaining two models renamed 240 and 340. For two more years they soldiered on, but with a cheapened specification, offering plastic upholstery and thinner bumpers to replace the Mk II's gleaming chrome double-deck fixtures.

Facts & Figures: The wilderness years

The Mk II's status as a collector's gem with a 50-grand price tag was hard-won. Thanks to the primitive monocoque construction inherited from the 2.4, there are nooks and crannies aplenty where water, mud, fungus and condensation can kickstart rust. Instead of being restored, in the 1970s many were simply entered into banger races – a sad fate for one of Britain's finest sports saloons with its gorgeous twin-cam engine and walnut-and-leather interior. It was completely rediscovered in the cash-rich 1980s (helped by its starring role in TV's *Inspector Morse*) and enough of the 83,000 built survived to provide a renovated Mk II for every city trader who fancied one.

THE JAGUAR E-TYPE

The Jaguar E-type was, and remains, a phenomenal car, an effervescent symbol of 1960s freedom – the ultimate contemporary performance car with a bird-pulling image at a time when bird-pulling was perfectly acceptable. If you'd made it, you drove an E-type, and you basked in the slack-jawed attention that followed it, and you, everywhere you went. 'The greatest crumpet-catcher known to man' was how Henry Manney, one roguish US journalist, framed it, although few owners today would dare admit to being swayed by the E-type's phallic powers.

The impact of the E-type's unveiling at the 1961 Geneva motor show was huge; Jaguar took 500 orders at this one event alone. Its aggressive, dart-like profile, with its lengthy, bulging bonnet and tapering, pointed tail were like nothing else on the road. Jaguar's streamlining guru Malcolm Sayer had taken the sleek contours of the D-type that won Le Mans in 1955, '56 and '57 and made sure they worked for a road car, while adding necessities like bumpers

ON THE MONEY

Price wise, the original E-type was an absolute steal. The roadster cost £2,097 15*s* 10*d* on the road. The Aston Martin DB4 cost £4,084 11*s* 5*d* but could only manage 140mph, while a Ferrari 250GT was megabucks at £6,469. By contrast, a small family car like the Triumph Herald cost £708, and an average house about £2,700. The V12 roadster repeated the bargain offer at £3,139 in 1972, when a slower, bulkier Mercedes-Benz 350SL cost a thumping £5,843. Production ended in 1974 after 72,007 E-types had been built.

Frank Sinatra is reputed to have barked, 'I want that car and I want it now' – here's why.

Facts & Figures

150mph or bust

Jaguar was anxious nothing should dent the 150mph E-type's mystique. Demo cars loaned to magazines offered blistering performance from the 3.8-litre straight-six motor, with 149mph attained for the roadster and the magic 150 for the coupé. But those cars had been illicitly souped-up: any E-type owner trying to reach a ton-and-a-half, maybe on the 72-mile-long M1 which had no speed limit until 1965, found his car ran out of puff at 140. . .

The Series III E-type was the first Jag to pack a spectacular V12 engine.

and a proper hood (even a practical hatchback third door on the coupé). Meanwhile, the stiff monocoque construction and supple independent suspension gave a fabulous ride and great roadholding.

Owners overlooked the E-type's cranky four-speed gearbox and four-wheel disc brakes that required a he-man shove to work properly, and simply roared off – grinning. Jaguar, however, soon set about improving the car. A bigger 4.2-litre engine from 1964 had more torque and came with vastly improved gearbox, brakes and seats. Collectors have long rated this Series I 4.2 E-type the cream of the crop. Later developments blunted image and performance; the 2+2 of 1966 looked ungainly, while the Series II models two years later lost the sleek headlight fairings, featured bigger bumpers, and had detuned engines to meet California's stiff pollution laws.

The E-type received a late-life fillip in 1971 when Jaguar installed its magnificent V12 engine, turning the Series III E-type into a powerful and unusually silent tourer that was, finally, a genuine (see side story!) 150mph machine.

Sir William Lyons introduces the E-type to an astonished press corps in March 1961.

THE JAGUAR XJ6

Jaguar launched the XJ6 and promptly rewrote the luxury car rulebook. Not with anything radical, mind: the XJ6 was a conventional saloon, front-engined, rear-driven, coil-sprung. They did it by masterful fine-tuning.

Jaguar combined in one design standards of ride comfort, silence, handling and roadholding – qualities previously thought incompatible in a luxury car – that eclipsed Europe's best and set the pace for the next 20 years. On its plump tyres, specially designed by Dunlop, this new British world-beater could out-corner Jaguar's own E-type but had a silkier ride than a Roller.

It had beauty too, feline aggression and organic muscularity. The company thought it so unmistakably a Jaguar that early cars didn't feature a Jaguar nameplate. 'A new yardstick' said *Autocar* magazine in June 1969. 'If Jaguar was to double the price of the XJ6 and bill it as the best car in the world, we would be right there behind them.'

Initially using the proven six-cylinder 4.2-litre XK engine, most XJ6s were automatic, all had power

No need to put the word 'JAGUAR' anywhere on the XJ6 – one look was enough.

THE XJ40

The original XJ6 was hard to better and its replacement, codenamed XJ40, was a long time coming. When it did arrive in 1986, it was with a new range of 2.9- and 3.6-litre AJ6 engines to replace the venerable XK. The angular shape and rectangular headlights were, perhaps, an uneasy break with Jaguar tradition, but buyers weren't deterred, as annual production reached an XJ6 peak in 1989 at 39,000 cars.

steering, and Jaguar built a few with a short-stroke 2.8-litre engine to beat European tax laws.

Like all previous Jags, the XJ6 was a bargain, often undercutting comparable Mercs by 50 per cent. From day one, buyers jostled to get theirs. An angry group from Switzerland even staged a protest to complain about the waiting list, while some second-hand cars changed hands at over list price.

The XJ6 wasn't perfect, of course. Quality, never totally exemplary, had tanked when the face-lifted Series III arrived in 1979. Then, the future seemed bleak for the still publicly owned, strike-torn Jaguar. Customers were in mutiny. Enter, in March 1980, John Egan, appointed by Margaret Thatcher to revive British Leyland's flagship. As his quality drive took hold, flagging sales were arrested, and Jaguar went public with spectacular success in 1984.

The final original-shape XJ (actually an XJ12) was built in 1993. Because of its umbilical link to the visionary William Lyons, to many it was the last proper Jag saloon.

Facts & Figures

The XJ12

Jaguar had hoped to launch the XJ saloon with its all-new V12 engine, but design fettling meant this awesome power unit didn't appear until 1972. At a stroke, this XJ12 became the planet's fastest production saloon car, at 147mph, and the only four-door V12. 'A marvellous achievement, deservedly the envy of the world,' declared *Autocar* in March 1973. The only transmission was automatic, and it was also the first Jag with standard air-conditioning. The most sumptuous edition was the Daimler Double-Six Vanden Plas with long wheelbase and individually contoured rear seats.

THE JAGUAR XK

The second generation of Jaguar XK has really put the lead back into Jaguar's pencil. It's a focused high-performance tourer, with two small seats squeezed in behind driver and passenger for small children or creased-up adults, and packing a phenomenal 510bhp punch from Jaguar's new, supercharged 5-litre V8 engine (in the XK-R). Top speed is electronically limited to 155mph, but this car could go much faster.

The XK employs cutting-edge technology. The structure is an all-aluminium monocoque, entirely bonded and riveted together, developed from aerospace-inspired techniques showcased in Jaguar's 2005 Advanced Lightweight Aluminium concept car. Both XK Coupe, at 1,595kg, and Convertible, at 1,635kg, are extremely light and torsionally rigid.

There's also a normally aspirated 5-litre V8 with 385bhp that, at 5.2 seconds, is only 0.6 of a second slower in dashing from 0–60mph. All XKs have a ZF six-speed automatic gearshift, controllable by either the JaguarDrive selector, new for 2010, or paddles on the steering wheel. An electronic differential is fitted to the XK-R to temper its blinding through-the-ratios acceleration; after all, it only takes 1.9 seconds to surge from 50–70mph. Yet the XK is careful with fuel, the XK-R managing 23mpg and the standard XK 25mpg. The performance is one half of the XK's distinctively British charisma. The styling, by ex-Aston man Ian Callum, is the other. It's a wonderful looking car, with just a hint of E-type to its grille. The leather-lined cockpit exudes a special aura, with its real wood veneers, while the power-top Convertible has a traditional fabric roof folding flush with the body contours for the ultimate in cruising style.

After requests, Jaguar's engineers retuned the exhaust system – not so it was quieter but so the throaty roar of the V8 was more audible inside the car. And Britain's walkers can feel more relaxed when they see an XK. It was the world's first car with a pop-up bonnet to cushion any soul unfortunate to make the wrong sort of contact with the bodywork.

◄ *Today's XKR packs 510 supercharged horsepower.*

Facts & Figures

The Jaguar XJ-S

Upholder of Jaguar's sporting traditions between 1975 and 1996 was the XJ-S, but it was a very different kettle of fish to the E-type. With the 5.3-litre V12 engine under its long, broad bonnet, it was an effortless performer, but as it was based on the XJ6 it was more refined grand tourer than energetic sports car. It was certainly distinctive, with those flying buttresses flowing out from behind the cosy 2+2 cabin, but buyers had to wait for eight years until an open-top version, and a less thirsty six-cylinder engine, was offered.

THE JAGUAR XK8

This suave, sophisticated GT and convertible replaced the aged XJ-S in 1996. Once again, it wasn't quite the 'new E-type' Jaguar aficionados constantly yearn for, but a comprehensive revamp of the XJ-S with new styling and Jaguar's 4-litre V8 engine. However, the XK-R added supercharging to the mix, turning it into a tyre-smoking fireball.

THE JENSEN INTERCEPTOR

The Interceptor was one of the most handsome British GT cars of the 1960s: and under its over-long bonnet was Detroit muscle car power. It was to enjoy a 10-year production life until, tragically, Jensen went bust.

By the 1960s, Jensen's C-V8 model was long in the tooth. It was very fast with its 325bhp Chrysler 6.3-litre V8 engine, but looked like a dog's breakfast, and sales were dwindling. So Italian design company Carrozzeria Touring was hired early in 1966 to sprinkle some design stardust on it.

Six months later, a running prototype was taken on a four-day, round-Italy road test. It fared so well that the car was shown at the Earl's Court motor show in October. It was mobbed by admirers and, at £3,743, orders poured in.

Despite having almost exactly the same chassis, the heavy, steel-panelled Interceptor could never be quicker than the glassfibre-shelled C-V8. Yet the Interceptor effortlessly managed 133mph and 0–60mph in 7.4 seconds. Anyway, relaxed cruising for business tycoons was the goal, not racing car performance for tearaways; a manual was theoretically available but the option was quickly ditched. There were four deeply padded, leather-clad seats and acres of luggage room under that distinctive, glass-domed tailgate.

THE AMAZING JENSEN FF

The Jensen FF – standing for Ferguson Formula – was a technological tour de force when revealed to an incredulous world in 1966. It was the first road car to combine four-wheel drive with anti-lock brakes. The Interceptor-based car had phenomenal traction and safe handling on slippery roads, as well as being fast and elegant. The FF was very expensive, its £5,340 sticker resulting in just 320 sales, but well ahead of its time: 14 years later Audi copied the concept for its Quattro, rendering the FF another great British idea that got away.

Facts & Figures: The Interceptor returns

From the ashes of Jensen's 1976 insolvency rose Jensen Parts & Service, established to ensure Jensen owners had continuity of spares and servicing. By 1983 this company had built a brand new Interceptor, the Series IV, and it was put into production, although priced at an outrageous £100,000. The firm was renamed Jensen Cars to reflect the rebirth. During 1993, however, even this company was in difficulties and closed down after selling barely a dozen cars. Several attempts have been made to revive Jensen subsequently, and all have failed.

The basic Interceptor remained unchanged throughout its life. There were Mk II and Mk III cars, a high-spec SP version and, in 1974 and 1976, elegant convertible and hardtop coupé versions respectively. The principal mechanical change was a hop-up from 6.3- to 7.2-litre capacity in 1973.

Jensen's liquidation in 1976 was down to the combination of a government-imposed three-day week, withdrawal of credit by Bank of America, and an ugly nadir in workforce/management relations. But the Interceptor lived on in the metal (see panel) but also in spirit as Jensen's contribution to the pantheon of desirable classic cars.

THE JOWETT JAVELIN

'A car of unusual merit,' gushed *The Autocar* magazine in June 1951, 'outstandingly good . . . a shining example of the better kind of family saloon . . . It is comfortable, and the quality of the motoring provided is very high as regards both engine behaviour and the riding and handling.'

It was describing the most advanced British family car of its time – all the more remarkable for the company that made it. Yorkshire brothers Benjamin and William Jowett introduced their own cars in 1913, a series of sturdy and basic twin-cylinder jobs. By the late 1930s, even the most austere Yorkshire owners were tiring of them.

Facts & Figures

Gerald Palmer

The creator of the Javelin, Gerald Palmer, built his first car – a Ford Model T with a plywood body – as a 13-year-old schoolboy in Rhodesia, where his father was chief engineer on the country's railways. While running the drawing office at MG in 1942, he couldn't resist Jowett's recruitment advert in a trade magazine. It was every car designer's dream: a blank piece of paper on which to create a 'six-passenger family utility': the Javelin. 'I did feel a little ashamed at times,' he admitted later, 'thinking of my career when there was a war on.' He returned to MG and later joined Vauxhall.

▲ *All Javelin passengers sat within the wheelbase, for comfort.*

◄ *The car looked ultra-modern in 1947.*

➤ *Somewhere below that radiator is the lively, flat-four cylinder engine.*

The Javelin, revealed to the public at the British motor industry's Golden Jubilee Parade in 1946, turned the company's rudimentary image on its head. Here was a potential British world-beater. For one thing, it featured unitary body/chassis construction. Then there was supple torsion bar suspension all round, independent at the front, pin-sharp rack-and-pinion steering, and a passenger compartment in which rear occupants sat within the wheelbase, helping excellent ride comfort.

The traditional Jowett flat twin engine had been redesigned, adding two extra cylinders and enlarging it to 1,486cc, and it was mated to a four-speed gearbox. And the whole thing was clothed in a wind-cheating teardrop-shaped body. Because it was light, aerodynamic and high-geared, the Javelin could belt along at 80mph, with excellent acceleration.

THE JOWETT JUPITER

Jowett was an unlikely entrant in the sports car arena in 1949 with the Jupiter. In an imaginative move, it engaged Dr Eberan von Eberhorst – an engineer who had worked for the mighty Auto Union grand prix team in Germany in the 1930s – to create a tubular spaceframe chassis for the car, which otherwise used Jupiter parts. The three-seater roadster was a sprightly performer with long-distance guts; in the Le Mans 24-hour endurance race, Jupiters achieved a hat-trick of class wins in 1950–2, including 13th overall in '52.

Owners loved the cars when they were new – price: £819 – but elation quickly evaporated as engines and gearboxes failed, or sometimes seized. This was tolerable in Britain but turned the car's reputation sour in far-flung export markets. The end came in 1954 when the company closed down, crippled by sub-standard, and delayed, component supplies. Today, the old Bradford factory is a Morrisons supermarket.

THE LAGONDA V12

After the bankrupt Bentley was rescued by Rolls-Royce, the towering figure of W.O. Bentley was retained to supervise a new range of sporting cars, bearing his name but based on Rolls components. You can imagine the ignominy. So in 1935, as soon as he was released from that obligation, he was delighted to join Lagonda. This marque alone made cars his way – without compromise. Indeed, Rolls-Royce dearly wanted to buy Lagonda, probably to eliminate the biggest threat to its Bentley division.

WO's influence was soon evident across the entire Lagonda range and, in 1936, the wraps came off what most people consider to be his masterpiece: the Lagonda V12 engine. It had a much shorter stroke than contemporary convention and each bank of cylinders had a single overhead camshaft. Even in standard tune, this gave 157bhp and the 4,480cc engine was installed in a new chassis with torsion bar independent front suspension and cutting-edge technology like dual-circuit braking.

Since it was intended to square up to Rolls-Royce, two years were spent in careful development so that, when the V12 began manufacture in 1938, it was as near-perfect as possible. With prices starting at £1,285 for a tourer with handsome factory-fitted coachwork, production would never be high, and only 189 were made before the Second World War.

In 1939, two lightweight V12s were entered at Le Mans, ran absolutely faultlessly, and finished third and fourth overall. They were carefully paced since the

▲ *Post-war Lagondas, like this 1953–57 3-litre, had to get by on a mere six cylinders.*

▲ *This V12 lit up Le Mans in 1939.*

Facts & Figures

Cross the road to take a look

The Lagonda V12 caused controversy at the 1937 motor show, the first held at Earl's Court in West London. Transport minister Sir Leslie Hore-Belisha – the self-same individual for whom the flashing pedestrian crossing beacon is named – turned puce at the sight of the car's streamlined horn fairings, thundering that their resemblance to female breasts was an affront to British morals. Lagonda staff made amends by removing the chrome horn outlet grilles that Hore-Belisha had found offendingly nipple-like, and covering the car's modesty with a front bumper.

rationale was merely to gain experience for an all-out attempt at victory in 1940 – a motor racing fixture, of course, that never took place. Existing V12 owners must have felt swelled-up with pride, though, as even most roadgoing versions could make 105mph. For the time, that was excellent going.

Lagonda's post-war owner, David Brown, decided the prevailing economic landscape was too austere for the V12. He was probably right. The irony is, though, Brown later did try to revive Lagonda's pre-war pinnacle with a new V12 engine, and it was a flop.

POST-WAR LAGONDAS

Lagonda, based in Staines, Middlesex, was acquired by David Brown in 1947 and merged with its near neighbour, Aston Martin of Feltham. It was Lagonda's 2.6-litre engine and Aston's chassis that formed the basis of the new Aston Martin DB2, but luxurious Lagonda saloons were on offer continuously until 1965 – the 2.6-litre, followed by the 3-litre and then the 4-litre Rapide. There was a brief revival in 1972 and then a proper one in 1976 with the infamous 'wedge' Lagonda.

THE LANCHESTER 38hp

Your country's first ever, all-British petrol-powered car with four wheels actually took to the road in December 1895, when 27-year-old Frederick Lanchester fired up his prototype and eased it forward for those first, historic yards on a gloomy day in Birmingham. Nothing, not even a man with a red flag, was going to stop him.

He had conceived his vehicle from the beginning as an automobile, and not some kind of horseless carriage. It featured cantilever spring suspension and a torsionally stiff chassis giving – for the times – an astounding ride. At first, a single-cylinder, 1.3-litre engine was fitted, the piston having two connecting shafts to separate crankshafts and flywheels rotating in opposite directions. This balance made for extremely smooth running. A two-cylinder engine was installed in 1897, a second complete car was constructed, and the Lanchester Engine Company cautiously inched towards production. Six demonstrators were built in 1900, featuring two-cylinder, 4-litre, horizontal air-cooled engines, retaining the twin crankshaft design. The transmission

Facts & Figures: The Unholy Trinity

Assisted by his brothers George and Frank (people mockingly called them 'the Unholy Trinity' as they worked on their cars on Sundays), Frederick Lanchester manufactured cars in Sparkbrook, Birmingham. He quit the business in 1913 to become a consultant engineer-inventor, lodging an incredible 400 patents, until his death in 1946, for everything from disc brakes to turbochargers. George then took charge of design, while Frank ran the London sales office. Despite a period in the 1920s when a Lanchester was many a playboy's dream machine, the Depression forced the company into a takeover by Daimler in 1931. Today, there's a stirring monument to the 1895 Lanchester car, by sculptor Tom Tolkien, in Birmingham's Bloomsbury district. Go and see it and pay homage to this marvellous British pioneer. . .

used epicyclic gears, steering was by tiller rather than steering wheel, and, despite being prototypes, the fastidious Fred ensured all components were completely interchangeable.

Lanchesters finally went on sale in 1901. Even then, as standard car design was materialising (the 1903 Mercedes 35hp set the pattern most would adopt), they were quirky. They had a hugely long wheelbase for comfort, as passengers sat within it, and a low centre of gravity as the engine was mid-mounted. The driver sat well forward, as there was no bonnet. Protection from the wind was provided by an apron like a large metal fig leaf. And, of course, until 1911 – when this 38hp model appeared – only tiller steering was available.

The 38hp is the acme of early Lanchesters. With its 4.8-litre, overhead-valve, six-cylinder engine – now water-cooled and featuring a pressurised lubrication system – it was the most refined and advanced British car. Lanchesters had a fantastic reputation among an informed élite, including best-selling authors Rudyard Kipling and George Bernard Shaw.

LOW-KEY DEMISE

Lanchester went out on a whimper in 1956, its pioneering currency having been systematically destroyed after Daimler stepped in and began attaching the Lanchester name to versions of BSAs and Daimlers. The last Lanchester, the Sprite, appeared more promising; it was intended as a compact automatic luxury saloon, but the project was axed by cash-strapped Daimler after a mere 10 had been made.

The innovation-packed Lanchester 38hp was unlike any other luxury car on the road. (Photograph courtesy of Chris Clark from his book The Lanchester Legacy *– www.lanchesters.com)*

THE LAND ROVER SERIES 1

Britain's government ruthlessly apportioned raw steel supplies after the Second World War to companies producing exportable goods to earn foreign currency. The brothers who controlled Rover, Maurice and Spencer Wilks, decided to bypass this edict.

They planned a four-wheel drive, tractor/pick-up hybrid tilted at farmers and made from war-surplus materials, principally aircraft-grade aluminium, to get around steel restrictions. Willys Jeep axles and body frames were used as a template but Maurice Wilks went for a single, central driving seat like a tractor. 'It must be along the lines of the Willys Jeep,' Wilks instructed his engineers, 'but much more versatile, more useful as a power source, be able to do everything.'

Once the Land Rover got the go-ahead in September 1947, the 80in wheelbase was the only Jeep connection, as most components came from the old Rover 60 saloon, including the 1,595cc four-cylinder engine and axles. The only bespoke item required was a 4x4 power transfer case. Good sense prevailed so that two seats and a normal offset driving position were provided, and the bodywork was simplified so Rover workers could form the panels using low-cost jigs and hand tools.

It was launched at the Amsterdam motor show in January 1948 at a mere £450, albeit to an extremely basic specification. When it went on sale in July that year, there was just one paint colour, Avro green – more war-surplus material!

With a 50bhp engine and a healthy 80lb/ft of torque at 2,000rpm, the Land Rover bounded nimbly up and down slippery hills on its Avon Trackgrip tyres. Excellent wheel articulation thanks to tough leaf-spring suspension and athletic

One of the earliest Series I Land Rovers in all its wonderful simplicity.

The boxy body was made of rustproof aluminium, from Army surplus materials.

A Series I doing its rural stuff.

EXASPERATED
AUNTIE

The Land Rover caused consternation at the British Broadcasting Corporation. With its lofty ban on mention of brand names, the presence of a Land Rover on a battlefield or royal parade presented a problem for newsreaders and reporters. No convenient 'generic' name existed so BBC wordsmiths invented one, 'field car'. It was used until the 1970s when 'Land Rover' was finally deemed acceptable!

approach/departure angles of 45/35 degrees were real assets. Four-wheel drive was permanent and, with no central differential and a freewheel in the front drive to reduce tyre scrub, slightly uncoordinated. In 1950, an ingenious dogleg clutch fixed that, permitting selectable two- or four-wheel drive; the definitive Land Rover 'system' had arrived. Pulling power mattered, not acceleration, so the tardy 0–40mph time of 18 seconds was irrelevant, and it offered 24mpg, something farmers applauded.

Rover envisaged selling 50 weekly before enough steel could be obtained to ramp up traditional car manufacture. Yet within a year, Land Rovers were outselling Rover cars, setting the pattern for an amazing future.

Series II to Defender

The Series Is saw upgraded engines and wheelbase throughout their nine years on sale; 200,000 were made, 70 per cent exported. The hugely improved Series II arrived in 1958, the Series III took over in 1971, and by 1976 production topped one million. The 90 and 110 arrived in 1984, adopting the Range Rover's coil-spring suspension, and they were redesignated the Defender range six years hence. It's great to report that, two decades later, they're as popular and versatile as ever.

A Series III Landie.

THE LAND ROVER DISCOVERY

Throughout the explosion of four-wheel drive during the 1980s, the craze was dominated by US and Japanese products. Land Rover, purveyor of indomitable transport to the world's armies and maker of the exclusive Range Rover, barely figured. Until, that is, 1989 and the launch of the Discovery.

With Japanese companies beginning to manufacture conventional cars in Britain, there was space left in their European import quota allocations for the Toyota Land Cruiser, Isuzu Trooper, Mitsubishi Shogun and their off-road ilk. These all offered good quality Range Rover alternatives for vastly less money, so Land Rover had to respond.

Discovery underpinnings were familiar and reassuring. It shared the Range Rover's construction method of separate chassis and mostly aluminium body panels.

Land Rover's design studio gave the Discovery an individual character, with a roof panel kicked-up above the rear quarter window. To make the car more spacious and family-friendly, the spare wheel was hung on the outside of the side-opening tailgate.

The interior benefited from lateral thinking thanks to the Conran Design consultancy; hence neat touches like a portable shoulder bag between the two front seats. The use of light blue and green colours, in place of the usual grey, gave the cabin a lift too.

At £15,750, it was half the Range Rover's price, and cost the same with a 2.5-litre turbodiesel or a 3.5-

➤ *In either three- or five-door form, the Discovery shook up the market for capable off-road vehicles.*

litre V8 petrol. In 1990, a five-door joined the range, and optional, rear-facing children's seats at the back turned the Discovery into a seven-seater people carrier.

The Discovery fairly blew its Japanese opposition into the weeds when it came to sheer off-road prowess. The uniquely capable permanent four-wheel drive, stout separate chassis and supple springing made it unbeatable. But the original Discovery also drove straight into Britain's suburban heartlands, establishing Land Rovers as plausible alternatives to traditional large family estates.

The Discovery offered Range Rover-class off-road ability, with a seven-seat option.

WHAT'S IN A NAME?

Land Rover went for a tongue-twisting, four-syllable title for the Discovery; the usual motor industry practice is to choose something blandly, inoffensively international, throwing up an automotive Esperanto stuffed with meaningless words like Zafira and Berlingo. Discovery was selected by the Interbrand consultancy. 'We wanted to find something that looked rugged both on the page and as you said it,' recalled Interbrand deputy chairman Tom Blackett.

Facts & Figures

Discovery II and III

This top-selling Land Rover gained a comprehensive styling overhaul in 1998. The original Discovery's proportions were retained except for a longer rear overhang; not so good for rock-hopping but the third row of seats could now face forwards. A bold new Discovery III was launched in 2004. It heralded new semi-monocoque construction, with the single-structure engine bay/passenger cabin mounted on a lightweight ladder-frame chassis for less lorry-like driving characteristics. Fully independent air suspension adjustable to driving terrain, and three separate electronic handling control systems improved everyday performance while keeping Land Rover's legendary off-road credentials intact. A new engine line-up features a 2.7-litre V6 diesel and Jaguar petrol V8s.

THE LOTUS ELAN

Until the Elan, Colin Chapman's Lotus road cars had been specialised affairs. Even the beautiful Elite of 1957, with its novel glassfibre monocoque bodywork, suited only the most dedicated enthusiast.

The Elan, then, was simpler. But it was even more brilliant. Its 'backbone' chassis, made of pressed steel, splayed out at each end, catapult-style. The front fork cradled the engine, the rear one the driven axle and differential, while both carried suspension units. The glassfibre bodyshell was delightful-looking, the only gimmick being pop-up headlights. Lotus's own, Ford-based, twin-cam 1,588cc engine, with 105bhp, gave the Elan 0–60mph acceleration of 9 seconds and a 110mph top speed.

The bare stats and details were one thing; it was the handling and roadholding that stunned. It was nothing short of sensational and Lotus owners soon became infatuated by the car's peerless road manners. In 1971, *Motor* magazine said of the by-then nine-year-old car: 'The Elan Sprint will go round most corners faster than visibility permits.' Meanwhile, Diana Rigg, as Emma Peel, endowed the Elan with karate-chopping sex-appeal when she drove one on TV in *The Avengers*.

The delightful Elan, a pure sports car whose peerless road manners inspired the Mazda MX-5.

THE 'NEW ELANS'

The Elan of 1989 was conceptually the opposite of its legendary 1960s namesake. The 1.6-litre Isuzu engine, usually turbocharged, drove the front wheels but despite the outstanding handling – once memorably described as 'tediously good' – it was never much-liked, not least for its strange looks. Meanwhile, the 1987 Evante was a loving update, and doppelganger, of the original Elan, with some specialist appeal until its maker went belly-up. But the real heir to the Elan's mantle was the Mazda MX-5 of 1989, which drew its inspiration directly from the dinky 1960s Lotus.

As the Elan matured, so it improved: the 1964 S2 replaced disc wheels with centre-lock ones, and was mostly sold as a hardtop; the 1966 S3 got a 115bhp SE version with close-ratio gearbox and servo-assisted brakes; the 1968 S4 added wider wheel arches surrounding low-profile tyres. Then for ultimate thrills the 1970 Sprint with its 126bhp 'big-valve' engine made 120mph and 0–60mph to 6.7 seconds possible, and looked fantastic in its two-tone paintwork.

The Lotus Elan +2

When Colin Chapman found himself the head of a young family, he realised the Elan needed to expand too if his customers weren't to desert him for other marques once their bundles of joy started to arrive. So the Elan +2 of 1968 was the one for the racy new dad, with two small rear seats packed into a wheelbase stretched by 30cm. It was wider too, and sold only as a coupé. Like all Lotuses, the beautiful Elan +2 had excellent ride and was easy on fuel – Chapman's light and aerodynamic cars were always efficient. From the start, the Elan +2 had the 115bhp engine; and the +2S 130 of 1971 had the 126bhp 'big valve' unit which meant 110mph and 0–60mph in 7.5 seconds, remarkable figures for a four-seat car with only a 1.6-litre engine.

THE LOTUS ESPRIT

It can drive underwater, Sharon Stone knows how to handle it, it's shaped like a ground-hugging missile, and it can top 160mph: it's the Lotus Esprit, the all-British supercar that – after 28 years on sale – is the longest-lived mid-engined two-seater sports car.

The Esprit was created when Lotus founder Colin Chapman decided to update his Europa – the first truly affordable, mid-engined sports car. He chose a young Italian he met by chance, Giorgetto Giugiaro, for the assignment.

The result was low-slung and ultra-wedge-shaped, a racing car for the road that you had to be agile to squeeze into. It was built around Lotus's famed 'backbone' chassis, with the engine centrally located, and the prototype caused a storm when unveiled at the 1972 Turin motor show (Giugiaro went on to design the Volkswagen Golf, Fiat Panda and many more).

It was on sale by 1976, at just over £10,000. Its few vices, like a claustrophobic cockpit and tendencies to overheat or vibrate, didn't stop well-heeled thrill-seekers from queueing up to buy one. For the Esprit had a top speed of 124mph, did 0–60mph in 8.4 seconds from a mere 2 litres, and it held the road like glue.

By 1980, and the launch of the 152mph Esprit Turbo, however, the Esprit was massively improved. In 1988 its sharp-edged looks were getting dated, so it was given

The original Esprit was a potent mixture of Lotus engineering brilliance and Giugiaro styling mastery.

▲ *The Esprit was built until 2004 – this is one of the last.*

Facts & Figures

Esprits on the big screen

The public beyond Lotus aficionados got to know the Esprit from the edge of their cinema seats. Colin Chapman persuaded James Bond film producers to feature his Esprit in the 1977 film *The Spy Who Loved Me*. Roger Moore's white Esprit SI was as fast at sea as it was on the road, for it was (on screen, at any rate) able to travel underwater. It featured concealed missiles, a periscope, or on-board radar. An Esprit also starred in 1979's *Moonraker* and *For Your Eyes Only* in 1981. Later on, the second-generation Esprit picked up its film career with appearances in *Basic Instinct* – driven, of course, by Ms Stone – *Pretty Woman* and *If Looks Could Kill*.

➤ *In* The Spy Who Loved Me.

EXCLUSIVITY

Despite its near three-decade stint in Lotus showrooms, the Esprit remained pretty exclusive, as only 10,400 were sold. By contrast, Porsche builds 100,000 cars a year, and even Ferrari cranks out 7,000. Famous Esprit owners have numbered Noel Edmonds, Mark Thatcher, racing drivers Mario Andretti and Ronnie Petersen, TV sports presenter Dickie Davies, Paul Newman, Nicolas Cage, Gary Rhodes and rugby star Victor Obugu.

a comprehensive redesign to freshen it up. Ten years later came the option of a 350bhp Lotus V8 engine, while the Esprit was also the testbed for Lotus's groundbreaking 'active suspension' research. Indeed, Lotus engineers experimented on the car incessantly – from 1989, there were seven different rear spoilers alone. Right to the end in 2004 it was assembled in its own small area of Lotus's factory at Hethel, near Norwich; with not a robot in sight, a dedicated team of 21 workers built two Esprits a week, each car absorbing 584 man hours.

THE LOTUS ELISE

Occasionally, a car explodes the very boundaries of its class. The Lotus Elise is one of those. By building its entire structure from extruded aluminium and then bonding, not welding, it together, Lotus created a small sports car with an unparalleled blend of lightness and strength. The bodywork was hand-laid glassfibre. Lotus could never afford to develop its own powerplant, so the first Elise in 1995 utilised the 118bhp 1.8-litre engine from Rover's MGF.

Then Lotus let its masterfully accomplished chassis engineers loose on it; they endowed the mid-engined Elise, weighing just 725kg, with such exquisite road manners even hard-bitten critics proclaimed they'd seen the future after 10 minutes at the wheel. In the Elise, Lotus created its finest sports car since the original Elan – the kind of clever, lightweight roadster that founder Colin Chapman always strove for. The Elise secured the Norfolk company's once shaky-looking future. Higher-performing and limited edition variations proliferated, a redesigned (entirely on computer, for the first time) Series II arrived in 2000, and by 2007 all Elises had moved to excellent, Lotus-customised Toyota engines. The 2011 Elise 1.6-litre has a CO_2/km emissions figure of 149g, making it the cleanest petrol-powered sports car on earth.

The Elise's rigid aluminium platform also supports the suspension, so it's a ready-made base for derivatives. In 2000, for example, the new Vauxhall VX220 roadster was an Elise derivative, built in Lotus's plant but using General Motors engines. Lotus's own spin-off models have included the luxury Europa, extreme 340R and track-biased 2-Eleven. It's underpinned the Dodge EV, Pininfarina Enjoy and Rinspeed sQuba concept cars and also – most famously – forms the basis of the much-talked about electric Tesla roadster.

The Vauxhall VX220 shares the Elise's aluminium innards.

The second-generation Elise.

THE LOTUS EVORA

Codenamed 'Project Eagle' before its unveiling in 2008, this Lotus comes with a unique boast. It's the only mid-engined production car with two-plus-two seating – a Lotus for the family man who can't bear losing his performance-driving fix despite new paternal responsibilities.

Mind you, the two extra seats can only accommodate small-ish kids. It has a boot, with built-in cooling system, just big enough for a folded baby buggy. Lotus has also designed a 'Plus-Zero' model, a strict two-seater with enough boot space for a bag of golf clubs.

At the central heart of the new dad-friendly Lotus is a 3.5-litre Toyota V6, a six-speed manual gearbox, and a bespoke AP Racing clutch. Suspension is double wishbone and power steering is standard to compensate for the extra weight the Evora carries over the Elise, as it is intended to be more sports-touring than sports-racing in its character. Nonetheless, *Autocar* magazine rated its 5.4 seconds to 60mph acceleration and 162mph top speed highly, crowning it Britain's best driver's car for 2009. It said: 'On the road – and on most tracks – you won't want more power,' going on to declare it: 'fluid, responsive, comfortable, controlled, fast and generally very accomplished indeed.'

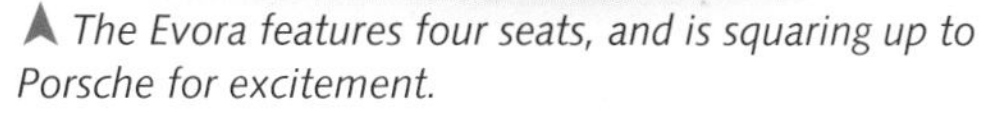

▲ *The Evora features four seats, and is squaring up to Porsche for excitement.*

Facts & Figures

Five-year plan; six-foot scale

The Evora is part of a Lotus plan to broaden the marque's appeal out from the core Elise, and really challenge Porsche. To make it a civilised GT to suit everyone, it was designed to accommodate drivers like Lotus's chief executive Mike Kimberley, who is 6ft 5in tall. Ironically, doctors ordered him to retire from the company in 2009, aged 70, due to a bad back.

THE LTI TX4 'LONDON TAXI'

As this book was being written, a sad state of 'a-fares' was reached in the long history of the Hackney Carriage or, as billions know it, the London taxi. Upstart Chinese carmaker Geely took majority control of struggling Manganese Bronze, the company that makes the LTI TX4. No sooner had they got their hands on this venerable British icon than they shifted manufacture to Shanghai. A handful of employees would remain in Coventry to assemble cabs from imported parts.

Nothing, it appears, is sacred in the car business. Then again, perhaps it's a measure of the TX4's unique built-for-purpose character that Geely can see the potential for the taxi in markets Manganese Bronze has never really cracked. With the recession knobbling cabbies' purchasing power, it only managed to shift 1,724 taxis in 2009, and a paltry 209 were exported.

We, the public, may love the upright TX4 – as familiar a London symbol as St Paul's Cathedral. Yet taxi drivers, traditionally the nation's biggest grumblers, increasingly prefer the comfort, easy maintenance and better build quality of modified Mercedes-Benz Vito minibuses.

▲ The TX4 is synonymous with life in London but has been acquired by the Chinese.

◄ The TX1 of 1997 was the biggest news in London taxis since the FX4, shown on the opposite page.

► This is the TX4 facet most of us are familiar with – the capacious passenger compartment.

Facts & Figures

The Austin FX4 & LTI Fairway

In November 1958 Austin set out to provide taxi drivers with a modern, easy-to-drive taxi: the FX4. It now had four doors, as the previous, antiquated FX3 came with three plus an open platform next to the driver on which luggage could be carried. Automatic transmission was standard, although a manual gearbox returned in 1961 after howls of protest from traditionalist drivers. It more suited the rackety 2.2-litre diesel engine (slammed in 1958 by the Noise Abatement Society, no less). It still adhered to rules, laid down in 1834, that defined 'Hackney Carriages', so there was enough rear headroom to accommodate a top-hatted gent. Sub-contractor Carbodies took over from Austin entirely in 1982. In 1989, it was renamed the Fairway and acquired a 2.7-litre Nissan diesel engine. It soldiered on until 1997 lodged in a time-warp, with those rear-hinged 'suicide' passenger doors, but well up to pounding London's mean streets.

The TX4 itself is a direct descendant of the Austin FX4, and so has a stout, truck-like separate chassis and a bought-in engine, a 2.5-litre VM Motori diesel yoked to a Jeep Cherokee automatic transmission. In this lineage, the biggest change came in 1997, when the TX1 cab updated the vehicle's styling, suspension (now coil spring instead of the old-fashioned cart springs) safety, comfort and accessibility. All TXs have a 25ft turning circle, to meet strict rules overseen by the Mayor of London so they can cope with narrow streets.

With Chinese ownership and Mercedes rivalry, the comforting sight of the traditional London taxi, preferably with its orange light signifying it's for hire, may not survive unchanged forever. Hail it while you still can!

ESSENTIAL KNOWLEDGE

All London taxi drivers – whether 'mushers' (owners) or 'journeymen' (renters) – spend unpaid years on mopeds, memorising London's streetscape to gain 'The Knowledge'. They must then pass an exam to prove they're capable of taking passengers anywhere within a statutory 6-mile radius of Charing Cross station . . . which *is* supposed to include south of the river at night-time.

THE MARCOS GT

The name is a compound of MARsh and COStin; Jem Marsh the car accessories retailer, and Frank Costin the aviation aerodynamicist Marsh commissioned to build him a small, race-winning GT car in 1959.

Costin was famed for his wind-cheating Lotus, Lister and Vanwall racing car bodies but, as a structural engineer, he also had vast experience of wooden military gliders. That's how he came up with a super-light monocoque chassis made from bonded marine plywood for the Marcos, with Triumph Herald independent front suspension and a live rear axle on coil springs and located by a Panhard rod.

This produced a highly effective track car – a pimply Jackie Stewart raced one – but it was an ugly brute, despite its gullwing doors. Marsh, though, knew an excellent sports car was in the making. In 1962 he commissioned a brand new body from stylist brothers Dennis and Peter Adams, and the immortal, long-bonnet/short tail/reclined driving position Marcos look created a sophisticated and slinky GT shape that ran the E-type close for glamour.

Frank Costin's ingenious wooden chassis was retained until 1968, and was only replaced by a steel version to meet US construction rules. The initial engine

The slinky shape of the Marcos GT ran the E-type close for visual impact.

Facts & Figures

Ever the Phoenix?

In 2000, Marcos went bust a second time, after almost two decades of offering a bewildering parade of models, race cars and engines all derived from the 1963 original. Canadian IT millionaire Tony Stelliga picked up the pieces and Marcos was back in 2002. He poured a huge amount of effort into a new car, the 185mph Marcos TSO, with a Chevrolet Corvette V8 engine and a chassis developed by rally team Prodrive, but in October 2007 the company folded for the third time in its 47-year history. Marcos, the great survivor, is sure to return sometime soon. . .

Marcos was revived in the 1980s, embracing Rover V8 power but keeping those classic lines.

choice was Volvo's 1.8-litre four-cylinder, but many Marcoses bought by Brits from the Bradford on Avon factory in Wiltshire were sold as kits, and Ford four-cylinder or V6 power prevailed (a Volvo 3-litre straight-six was specified for America). A unique feature was a sliding pedal box, rather than an adjustable seat, to accommodate any size of driver, Jem Marsh being 6ft 4in. Over-ambitious expansion put Marcos into liquidation in 1971, but Marsh bought back the name in 1976 and five years later reintroduced the classic Marcos as a kit car. Before long, full production versions were offered again, with a convertible option and the fashionable 1980s choice of Rover V8 3.5- and 3.9-litre engines turning the new Marcos Mantula into 150mph slingshots. One thing that never changed, though, was the laid-back Marcos style – just 43in tall and still recognisably the same as it was in 1963.

THE MINI MARCOS

Jem Marsh never forgot the grass roots sports-racing enthusiast. While the Marcos GT was winning praise, Marsh unveiled the Mini Marcos in 1965, a tiny glassfibre fastback featuring cheap Mini components. You can still buy one today. In 1966, a Mini Marcos was the only British-made car to finish the Le Mans 24-Hour race; as the drivers were celebrating afterwards, though, the car was stolen and never seen again.

THE McLAREN F1

Gordon Murray, chief Formula 1 designer at McLaren, enjoyed the privilege of building this pure supercar, with no expense spared. The ultimate roadgoing machine of its day, the McLaren F1 remained the world's fastest production car for 10 unbroken years.

His vision called for the fastest, most involving road car ever, yet one docile enough to drive into the city centre. Murray announced his plans in 1990. He would head design and development, Lotus stylist Peter Stevens would pen the car's shape, and BMW Motorsport agreed to furnish a custom-made 6-litre V12 engine.

Murray's no-compromise ideals included the three-seater cabin with the driver sitting centrally ahead of two passengers, the mid-mounted engine, and the world-first use of a carbon-fibre composite monocoque. Weight was crucial to anticipated performance: an incredible 1,000kg (among rivals, Ferrari's F40 weighed 1,235kg, Jaguar's XJ220 1,470kg). With 627bhp on tap, the truly exceptional performance Murray craved was made real.

Formula 1 driver Jonathan Palmer drove one at Italy's Nardo test track in 1993 to an incredible 231mph. Tests showed more performance wonderment: 0–60mph in 3.2 seconds and 0–100mph in 6.3. 'The F1 is the finest driving machine yet built for use on the public road,' gasped *Autocar* magazine, after conducting the only full 'consumer' road test. 'It has the best engine, the finest chassis, and is faster above

➤ *Each McLaren F1 road car – and there were only 64 of them – took 6,000 hours to build.*

▲ *A prototype undergoing a final shakedown at the MIRA track in the West Midlands.*

170mph than a Formula 1 car.' It remains the world's fastest ever naturally aspirated production vehicle.

Such performance made the F1 invincible in its subsequent motor sport exploits. F1 GTRs came first in every GT Endurance race they entered bar two, while the F1 was triumphant at Le Mans in 1995. It's the last road car-based competitor to win the 24-hour event.

The price, £635,000, reflected the astronomical development and production costs and that each example took 6,000 man hours to build, with 3,000 hours spent on each carbon-fibre tub. McLaren sold 64 road cars, all at a loss, out of a 107 total that included prototypes and racers. Still, one recently fetched £2.35m at auction, so as investments these cars are as gold-plated as their Facom tool kits.

Facts & Figures

Carbon-fibre evangelists

McLaren pioneered this strong, light and safe composite material in Formula 1 construction on the MP4/1 in 1981, Ron Dennis's first Formula 1 car at McLaren; previously it was used only in the aerospace industry. Its impact in motor racing was instant when John Watson's car was sliced in two at the 1981 Italian Grand Prix, after he went off the track at 140mph. The carbon-fibre structure remained intact even as the engine and transmission were torn off, and Watson walked away unscathed. Its strength-to-weight ratio far outperforms aluminium alloy structures, and all McLaren's competitors were immediately compelled to follow suit.

THE MERCEDES-BENZ SLR McLAREN

The SLR was conceived and styled by Mercedes-Benz as a powerful touring sports car before being presented to McLaren Automotive to engineer, develop and manufacture. The SLR was prodigiously fast (208mph is possible), exclusive, and a technological tour de force. With 2,252 examples produced up to December 2009, the SLR was an outstandingly lucrative supercar for all involved.

THE McLAREN MP4-12C

Relatively small British company decides to design, manufacture, market and look after a supercar to rival twenty-first-century Ferraris; followers of this country's ever-optimistic sports car industry may feel they've heard such ambitions before. Only, this time, the enterprise is McLaren and, considering its record in Formula 1 – eight constructor's championships since 1966, winning one in four races it's entered – the credentials are far removed from the typical British have-shed/will-tinker designer who makes good reading but, rarely, good cars.

The unveiling, in early 2010, of not just a new car but a brand new supercar organisation.

The £175,000 MP4-12C has the Ferrari 458, Lamborghini Gallardo, Porsche 911 Turbo and Aston DB9 in its sights. It's a mid-engined two-seater built around a carbon-fibre composite central tub called a MonoCell; it weighs just 80kg (of a total 1,300kg) so a sensationally nimble and secure driving experience is in prospect. The chassis is so strong the ultra-thin body panels play no structural part, their function purely aerodynamic.

Wing-style doors reveal a driver-focused interior bristling with new thinking.

McLaren has created its own, compact 3.8-litre twin-turbo V8 engine. It will propel the car to 124mph in under 10 seconds, while composite brakes will bring it to a halt from 100mph in 30m. The seven-speed transmission is a Seamless Shift dual-clutch. Derived from F1 cars, it allows pull/push gearchanges with either hand as the rocker-mounted control pivots with the

Facts & Figures: A brand new sports car company

McLaren Automotive has been planned from scratch, not only to make up to 4,000 cars annually at its futuristic Woking complex, but also to ensure high-net-worth owners are subjected to only the best aftersales service. It's all awesomely professional in a sector that's often endearingly shambolic. Chairman Ron Dennis outlined his vision: *'We believe our own incredibly high demands will be a positive example of how the UK can design, engineer and build world-class innovative products. We hope to inspire future generations of designers and engineers to work in these fields in the UK. The team here will not rest until the range of McLaren high-performance sports cars is considered the best in the world.'*

Cover blown on the MP4-12C's inner secrets, such as the 80kg carbon-fibre Monocell and an all-new engine.

steering wheel. The system has a function called 'Pre-Cog', to speed up reactions for quicker, more stimulating changes when desired. Metal bulk has been cut by using a driver-adjustable pro-active control unit where electronic reactors have replaced anti-roll bars, and a limited-slip differential has been jettisoned for a system that subtly applies one rear brake to boost cornering grip and preventing slide-inducing understeer.

Gullwing doors, a rear airbrake, and optional camera monitors (that can record your own road movie!) are among the exterior talking points. Another is the car's efficiency: McLaren reckons its 600bhp engine will emit just 300g/km of CO_2, making it cleaner per-horsepower than any other road car, hybrids included.

Yep, Porsche and Ferrari should be alarmed…

ADVANCED TELEMATICS

Haptics – the 'look', 'feel' and 'touch' of a control – have had a profound influence on the interior of the MP4-12C. The 7in touch screen for the telematics system is a fascinating example, as it's been orientated in an upright 'portrait' mode, a first for the automotive industry, like a book or a mobile phone interface. It's also been designed with a pleasing minimum of command buttons so it's less distracting.

THE METROPOLITAN

Since 1945, Nash Motors of Kenosha, Wisconsin, had been mulling a really small 'sub-compact' car. It asked freelance Detroit designer William Flajole to work up some proposals. Market research was in its infancy, but Nash decided to consult the public in 1949, organising 'Surview' events at which potential customers were canvassed for their opinions of Flajole's drawings for NXI (Nash Experimental International).

The answer from the suburbs reflected America's changing demographics: it must be cheap as well as dainty because an NX1 would be used mostly as a second car for shopping. Nash was convinced a good market was ripe for exploiting, but had no experience in building small cars, no suitable components, and no spare factory space. So, rather than change the formula, they picked a European sub-contractor: Austin. It had the ideal 1.2-litre engine and three-speed gearbox, from the A40 Somerset, and was eager to build cars for US export. A textbook car industry joint venture emerged.

Production began in October 1953, and the Nash Metropolitan went on sale in March 1954 exclusively in the USA and Canada, with a marketing campaign aimed primarily at housewives. It came as a two-door convertible or hardtop, and a radio and heater were fitted to every car. In 1956, a larger 1.5-litre engine headlined detail and styling improvements, although top speed remained slothful, at just 75mph.

In April 1957 Austin launched the Metropolitan in Britain, where its vibrant two-tone colour schemes and transatlantic looks set it apart . . . as did its appalling turning circle and the general driving

USA market research results made the Metropolitan one for the ladies to do their shopping in.

ALMA'S METRO

Alma Cogan, the British pop singer, died of cancer aged just 34. At the time, in 1966, her own Metropolitan was in for repairs at the Rambler/Nash service depot in north-west London. Fourteen years later, when Honda UK bought the building, they found her treasured car on the top floor, a rusted hulk; it had been kept in lieu of the unpaid bill.

impression that all its faculties, from steering to cornering, were suffused with blancmange.

An externally opening boot lid belatedly appeared in 1959, its best American year, when 22,209 were sold. This made it the second best-selling import after the Volkswagen Beetle. Austin did excellent business by building 104,377 of them up to 1961, with 94,986 destined for US/Canadian export, so it was odd no replacement for this important British export success ever materialised.

➤ *This is the interior of a Series III Metropolitan, featuring fashionable 'hound's tooth' upholstery fabric and white vinyl trim.*

Facts & Figures: The Austin A40 Somerset

Another British car leading a double life abroad was the bulbous Somerset, an underpowered four-door family saloon on sale in the UK between 1952 and '54. Egged on by the Japanese government to establish a domestic motor industry, Nissan Motor Co. signed a deal with Austin in 1952 to build 2,000 Somersets from imported kits. Austin was chosen because it was then the biggest-selling imported make in the USA. The agreement stipulated the Somerset's Japanese content would increase as Nissan got the knack of carmaking; it began with locally made tyres, batteries and flat glass, but the endgame came in August 1956, when the later Austin A50 Cambridge became 100 per cent 'Made In Japan'.

THE MG 'OLD NUMBER 1'

Morris Garages, William Morris's Oxford-based retail chain, appointed a dynamic young salesman called Cecil Kimber in 1922 as general manager. He immediately injected a new vigour into the business, expanding its service of kitting out Morris cars with customised coachwork and equipment.

In 1924 Kimber introduced an 'M.G. Special Four-Seater Sports' based on the Morris-Oxford, with spiffing polished aluminium bodywork. Using this as a basis, a special trials car version was constructed at a cost of £279. The modified chassis featured taut semi-elliptic spring suspension, lowered steering, and a trick overhead-valve 1.5-litre engine. The rakish two-seater had tiny mudguards and no windscreen or hood.

▲ *Cecil 'Mr MG' Kimber at the wheel of what came to be known as 'Old Number 1'.*

Facts & Figures

The MG M-Type Midget

This was the first truly affordable MG, built at the company's dedicated factory at Abingdon from 1929 – a roadster version of the Morris Minor to rival Austin's two-seater Seven models. An overhead-camshaft, four-cylinder engine, with improved suspension and steering, enormously improved performance, and the £185 price put it within reach of even impecunious motorists.

Kimber attacked the Land's End Trial in it in 1925, and scooped a gold medal in the Light Car Class. The 80mph car attracted widespread attention, soon leading to these sporting Morris Garages products becoming known as 'MGs'.

After the trial, the car was sold for £300, but an MG employee found it in a Manchester scrapyard in 1932 and bought it back for £15. The company then restored it – changing the original grey paintwork to bright red – to use for publicity, whereupon the dubious title of 'Old Number 1' was adopted. You can see it today at the Heritage Motor Centre in Gaydon, Warwickshire.

THE MG TC

The MG TC was pretty old-fashioned, even for 1940s Britain. It shared much with the pre-war TB, including tall 19in wire wheels, a fold-flat windscreen, and a slab-style petrol tank mounted behind the seats. Its short-stroke engine provided 54bhp, so a 68mph top speed was hardly stunning, and its all-round semi-elliptic spring suspension gave a bone-shaking ride.

But none of this really mattered. It had the look and the responsiveness of an authentic sports car, despite its flexible, vintage-style chassis with beam front axle. To enjoy his TC, a driver had to relish its shortcomings, concentrate and really work the car . . . quite a revelation to US servicemen in Britain after the Second World War, who were more accustomed to merely holding the tiller on the bridgehead on large sedans back home.

A few soldiers and airmen took a TC home with them, where its sprightly performance enchanted their fellow countrymen. San Francisco entrepreneur Kjell Qvale decided to import a few in 1947, a venture viewed as so risky his bank refused to back it. Buyers, however, shared Qvale's enthusiasm for the TC's scintillating appeal, and it became a smash hit, despite being available only with right-hand drive.

The TC provided racing thrills to Sports Car Club of America members, and also laid the foundations for a 30-year American sales bonanza for British sports cars.

The TC was fully in the British sports car tradition, and quickly captivated America.

Facts & Figures

The MG TD

Introduced in 1949, the TD kept all the arm-out, wind-in-the-face fun of the TC with crucial improvements like independent front suspension, rack-and-pinion steering, and factory-tuned engines offering up to 90bhp. The one downside was pressed steel wheels inside of wires, but it was still immensely popular.

THE MGB

Throughout its 18-year life the MGB was an honest, enjoyable, sexy little roadster, with a snug-fitting cockpit, a growly exhaust note and entertaining road manners.

The B adopted unitary construction, giving more passenger and luggage room than the outgoing MGA. The MGA's B-Series engine was bored out from 1,492 to 1,798cc, generating a healthy 8bhp increase in power to 94bhp, and the MGA's proven front and rear suspensions were carried over intact.

Announced in October 1962, this two-seater looked and felt modern but was built on solid MG virtues. Comfort was an important aspect. It was easier to get into than the MGA and had winding (not sliding) windows. At £950 on the road, the MGB was an instant hit.

Performance was lively. Top speed was 103mph and 0–60mph was accomplished in 11.4 seconds. The engine was hearty, with plenty of torque, and satisfyingly predictable roadholding shifted to entertaining handling on twisty roads.

The MGB GT coupé arrived in 1965, a pretty car with its elegant, Pininfarina-styled fastback incorporating a rear hatch and minuscule rear seat. The extra bodywork carried a 220lb weight penalty, so acceleration was blunted, but the more aerodynamic shape pushed top speed to 106mph.

Under inept British Leyland stewardship from 1968, little was invested in the MGB, despite its sterling work as a top British export to the USA. To meet North American impact safety requirements, in 1975 all MGBs gained unsightly black polyurethane bumpers;

➤ A beautifully mown cricket pitch and an MGB roadster on wire wheels – two aspects of British life that are well worth cherishing.

to meet US headlamp regulations, the ride height was raised by 1.5in; to satisfy emissions laws (on USA-bound Bs), a single carburettor and a catalytic converter were imposed, delivering emasculated performance for American customers – its 62bhp of power gave 90mph tops, with 0–60mph taking a dismal 18.3 seconds. Despite all that, the MGB lasted until 1980, shifting an amazing 513,000. 'Hot hatchbacks' supplanted open two-seaters as affordable fun cars during the 1980s, but second-hand MGBs were constantly in demand.

Facts & Figures

The MGB GT V8

A polished attempt to pack the MGB with a bigger punch came in the MGB GT V8 in 1973. Rover's 137bhp 3.5-litre V8 was installed, good for 125mph and 0–60mph in 8.6 seconds. Moreover, the alloy engine weighed an amazing 40lb less than the B-Series iron lump, so handling was better. But it was short-lived, thanks to the unfolding 1970s fuel crisis.

THE MGA

The pretty MGA of 1955 was derived from a one-off MG TD raced at Le Mans in 1951 and was based on a substantial box section chassis. The handling of the A was more than a match for contemporary Triumph and Austin-Healey rivals. On 1,489cc and 72bhp from its B-Series engine, it wasn't wildly quick but 95mph was handy, as was the potential 30mpg. However, an attempt to sex-up the MGA in 1958 turned sour when the special twin overhead-camshaft engine intended to make it a blistering 110mph car developed a nasty reputation for cooking its pistons. The high-octane petrol it needed wasn't widely available. The Twin-Cam was quietly axed in 1960.

THE MGF

MG went from top-selling sports car to annihilation in 20 years under British Leyland stewardship; it was careless destruction of a great British asset – like the Peak District gradually being covered by a housing estate. And when a revived MGB (the RV8) appeared in 1992, it looked like a last gasp of motoring 'Heritage Britain'. Dedicated sports car fans had plenty of reason to cry into their warm beer. But then, in 1995, the fortunes of the affordable British two-seater were dramatically reversed: MG was back, and on top form.

➤ *The mid-engined MGF was a tremendous technical achievement that was also scintillating to drive.*

The MGF was the epitome of the modern two-seater roadster, and also an extremely advanced design with its mid-mounted engine and clever Hydragas suspension. The system employed interconnected fluid and gas displacers, which provided a surprisingly compliant ride but which could be tuned to give excellent handling characteristics. The inherent balance this layout creates provided outstanding roadholding with none of the nasty breakaway surprises of traditional rear-drive roadsters.

The engine itself was a 1.8-litre edition of Rover's tremendous 16-valve K-Series – all-aluminium, twin-cam, fully catalysed and giving 118bhp in the MGF 1.8i. The 1.8i VVC added Rover's pioneering variable-valve timing for 145bhp, enough to propel the MGF VVC to 60mph in 7.6 seconds, and on to a 131mph top whack.

The F was refreshingly neat and trim from every angle. Careful attention had been lavished on details to delight owners – a riveted chrome petrol cap, black-on-white instruments, the MG logo moulded into the dashboard top, and so on.

A cosmetic refresh in 1999 was accompanied by a budget 1.6-litre engine and automatic transmission options, plus an even more powerful, limited edition 160bhp model called the Trophy 160. These changes sustained the MGF's position as Britain's top-selling sports car, constantly fending off the challenge of the more conventional Mazda MX-5.

Facts & Figures

The MG TF

New ownership of MG heralded a new lease of life for the mid-engined two-seater. The MG TF was rather more than just a mere makeover: the Hydragas suspension was abandoned in favour of conventional (cheaper to make and maintain) coil springs. These allowed the car to sit lower on the road at all times, best to exploit a body now 20 per cent more torsionally stiff for even more limpet-like cornering. Production ended in 2005 as MG Rover was forced to call in the receivers.

MG IS DEAD. LONG LIVE MG!

Most of the old MG assets have been snapped up by Chinese bargain-hunters, led by Nanjing Automobile. MG saloons are now built in Shanghai and sold locally, where buyers are told that the initials stand for Modern Gentleman, which is no doubt easier to explain than 'Morris Garages'. But it isn't quite the end of MG in Britain. In 2008, MG TFs began to be assembled again in a corner of the old Longbridge works, built up from imported kits of parts. It's lightly updated inside and out, with perhaps the key improvement being an engine that can resist the old TF's main bugbear: blown engine head gaskets.

The Mini

The Mini's ingenious concept meant it offered staggering interior space for a 10ft-long car. It was the 24-month design mission of Sir Alec Issigonis: single-minded, erudite, and a chain-smoker.

In 1956, British drivers faced petrol rationing in the face of the Suez Crisis. Demand for economical cars was enormous, leading to booming sales of flimsy 'bubble' cars. Issigonis hated them. Instead, he envisaged an ultra-compact 'cube' for four passengers, headed by a space-saving front-wheel drive system. In a leap of imagination, he positioned the engine transversely across the car, with the gearbox underneath instead of behind it, to keep the drivetrain super-compact.

Every available space was utilised: there were tiny 10in wheels in tight wheel arches, a shelf instead of a dashboard, and novel storage bins inside the doors. There was nowhere to install a radio, but Issigonis did include an ashtray!

At £496, the Mini was virtually the cheapest new car on sale. Yet the British Motor Corporation (BMC) suffered for backing Issigonis's engineering genius: Ford's staff dismantled an early Mini and costed each component. It calculated – correctly – BMC was losing £30 on every Mini sold. Still, because of its huge technology investment, halting Mini manufacture would have cost BMC even more. The little car only started generating a decent profit in 1982.

CAUTIOUS WELCOME FOR AN ALL-TIME GREAT

Consumer wariness meant that, in 1960, only 116,000 Minis were sold, way below BMC's capacity, and it didn't reach its peak sales year until 1971, when 318,475 found buyers. By the time the final example was built on 4 October 2000, 5,387,862 had been produced, making it the most successful all-British car ever.

The Mini phenomenon was ignited by simple word-of-mouth, as owners discovered it offered something no economy car ever had before: sheer driving enjoyment. The roadholding was a revelation. With the surefootedness of front-wheel drive and the go-kart like sensation of having a wheel at each corner, the Mini's corner-taking ability was extraordinary. This nippiness made it swift from A to B; so the meagre power from its 848cc engine hardly mattered, while its fuel economy was a genuine boon.

The Mini became the must-have city car. Glamorous owners included Paul McCartney, Margot Fonteyn, Twiggy, Steve McQueen, Spike Milligan and Peter Sellers. Motor sporting heroes from Graham Hill to Enzo Ferrari loved them, and the word 'Mini' passed into everyday English parlance when Mini-owner Mary Quant created the eponymous skirt.

➤ *This sectioned Mini reveals the amazing interior space it boasted.*

◄◄ *The last classic Minis of 2000 were still basically the same as the 1959 originals.*

◄ *The Mini Cooper reappeared in 1990 due to huge pent-up demand, especially from Japan.*

Facts & Figures: The Mini Cooper

John Cooper, F1 car constructor and a friend of Issigonis's, recognised the Mini's extraordinary performance potential. BMC liked his proposal for an image-raising 'Mini Cooper': Cooper boosted capacity from 848 to 997cc, and power from 34 to 55bhp with twin carbs and a modified camshaft. A real hotrod, over 87mph was possible, fast enough to justify fitting tiny front disc brakes. Later came the Cooper S, ultimately giving 76bhp for 100mph and greased-lightning acceleration. BMC took it rallying with incredible success. The Cooper S won the legendary Monte Carlo Rally in 1964, 1965 and 1967, driven by the likes of Paddy Hopkirk and Timo Mäkinen. Axed in 1971, the Mini Cooper was revived in 1990.

THE MINI MOKE

Flat-panelled and more open-air than any sports car, the Moke – Australian slang for donkey – was adopted as Britain's eccentric riposte to America's beach buggy. As carefree beach transport around the Mediterranean, it was perfect, but many stayed in 1960s London, where Afghan-coated hippies shivered at the wheel at Notting Hill traffic lights. *Motoring Which?* summed up the 'groovy' experience: 'Driving through the back streets of Kensington in pouring rain in the Moke must rate very low on anyone's fun index.'

Yet Mokes were originally intended to be stacked, windscreens folded flat and wheels resting on the mudguards of the car below, for packing into military transport planes. After being parachuted into combat zones, the idea went, it would be light enough to be carried on the shoulders of four burly squaddies if driving conditions overwhelmed it.

British army chiefs trialled prototypes; their reaction was lukewarm. Its low ground clearance, tiny 10in wheels and two-wheel drive hindered progress over anything more arduous than wet grass. It could be hoicked across swampy ground, but extra heavy weaponry made carrying a Moke back-breaking. The Royal Navy bought a handful but the Army stuck to Land Rovers.

Instead, it was launched in 1964 as an Austin or Morris Mini Moke. Most were exported and sold as

➤ *A flimsy plastic top was barely capable of keeping the Moke draught-free but at least made it practical everyday.*

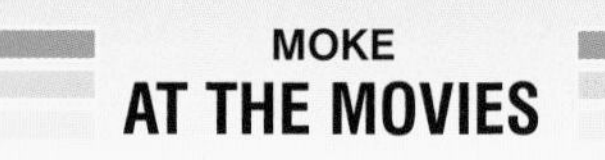

MOKE AT THE MOVIES

The Moke found a niche as an excellent film and TV prop, because cameramen could keep actors driving them engagingly 'in shot'. Mokes featured prominently in cult ATV series *The Prisoner*, and in the amazing underground scenes in *The Man With The Golden Gun*.

Facts & Figures: The Moke emigrates

The Moke tooling was shipped to Sydney in 1968 to begin a new life as Australia's cheapest domestic car; it was sold under the slogan 'Moking is not a Wealth Hazard' for A$1,295. Local changes numbered bigger 13in wheels and gutsier engines. Leyland even sold some to the Australian Army. By 1981, another 26,142 had been built there and then manufacture switched to Portugal, where 9,277 more were made until 1994. The Moke was used as a bargaining chip to sell Austin Metros there before full EC trade liberation – if you built cars in Portugal, the government granted a bigger import quota.

hotel taxis in hot countries. In Britain, the £405 price undercut every four-wheeled car on sale, for one simple reason: it wasn't actually a car. As the Moke came with just a driver's seat – and one windscreen wiper – Customs & Excise classified it as a commercial vehicle, on which no VAT was payable. It was slow, taking 22 seconds to hit 60mph, and super-basic, with an open-sided canvas tilt/hood, but the storage lockers built into its sides were useful. The sole colour option was dark green.

However, very few went to British window-cleaners and delivery men: once you'd bought your Moke, the trick was to turn it into a car by adding three optional passenger seats, an extra wiper and flimsy sidescreens. By 1967, the VAT man noticed the tax dodge and swiftly reclassified it as a car, raising the price by £78, and killing UK sales stone dead after a mere 1,467 had been sold here.

▲ *One of the many uses to which a Moke can be put – just don't send it on to a battlefield!*

THE MINI

For Britain, the 'classic' Mini became, simultaneously, a symbol of embarrassment and pride. We were ashamed of ourselves for never properly replacing it, but then glad it sailed on year after year needing no fundamental changes. Fortunately, BMW – which acquired the Mini in its takeover of Rover Group in 1994 – saw a way to fold the greatest small car of all time in with a vision of how to reinvent it for a new generation.

From the moment the wraps came off the first concept 'Mini Cooper' at the 1997 Frankfurt motor show, we realised BMW's approach was infused with a new purpose. Featuring front-wheel drive, power units fitted crosswise at the front, short body overhangs and ample space for four, the new models took up key elements of the original. It recaptured the cheeky charisma in a bigger car with an interior focusing on driver and passenger; no longer was the Mini aimed at hard-pressed young parents but carefree singletons and double-income/no-kids couples. There was, of course, a hatchback too, while the engine was a considerable step up to 1.4 or 1.6 litres.

The new MINI (BMW introduced these capital letters as it went on sale in 2000) was an opinion-splitter. Some bemoaned its reliance on a pastiche 'repositioning' exercise to revive Mini interest. Hardcore

Facts & Figures

The triangle of virtue

Since 2001, BMW has invested £400m in the MINI plant in Oxford – the Cowley factory originally opened in 1913 by William Morris, where 600,000 of the original Minis were made between 1959 and 1968. It's a cornerstone in a 'production triangle' of British plants contributing to the MINI: engines come from Hams Hall, Birmingham, body panels from Swindon, with painting and final assembly taking place at Cowley. Indeed, when the paint shop was under construction in 2000, it was Britain's second-largest construction project after the Millennium Dome . . . and has proved considerably more useful.

▲ *The compact MINI convertible has helped to make the USA the car's biggest worldwide market.*

From left to right, the MINI One, the Cooper and the Cooper S form the backbone of the range.

This 2002 line-up shows a fraction of the different liveries that could be ordered.

THE CLUBMAN

Clubman is the name of the MINI estate, introduced in 2008 and reviving the twin, van-like 'barn doors' reminiscent of the Mini Countryman and Traveller estates of the 1960s. There's more rear legroom thanks to a 3.15in wheelbase within a 9.45in overall stretch, and access to the family-friendly rear seats is via a third passenger 'Clubdoor' on the right-hand side of the car.

diehards decried the shift of focus from economy car to premium-priced lifestyle accessory. Punters, though, absolutely loved BMW's new baby. It was fun, nippy, individualistic, stylish and well-made. Annual sales immediately broke the 100,000 a year forecasts and have risen every year since launch, recession notwithstanding; 232,000 MINIs were sold in 2008, with the USA the biggest of 80 worldwide markets. Unlike in the complacent old days, new MINI has evolved continually: BMW rapidly added to the basic MINI One and sporty Mini Cooper models with a diesel option, a supercharged Cooper S and, in 2004, a dinky convertible. For the unreconstructed petrolhead there were 'John Cooper Works' specials, and in 2008 a longer, safer MINI MkII appeared; BMW kept the by-now cherished new MINI profile, although every single exterior panel was subtly updated.

THE MORGAN 4/4

Morgans have been handmade in Worcestershire since 1909, and the Morgan Motor Company has occupied its Pickersleigh Road factory at Malvern Link since 1919. Furthermore, it's been producing – at a steady, unhurried pace – the Morgan 4/4 since 1936. It's become the most venerable model name in motoring history, for a 4/4 has been on sale from then on.

Well, almost constantly: in 1950, the 4/4 took a breather while Morgan concentrated on bigger-engined cars. But for many customers, the joy of owning these uplifting sports cars came from wind-in-the-hair, vintage character and enjoyable, well-balanced handling; outright performance was secondary. So the 4/4 returned in 1955 with a trusty Ford sidevalve engine providing modest but adequate service. The recipe has remained intact ever since: the very latest 4/4 comes with Ford's four-cylinder Zetec, a 1.8-litre, 16-valve engine delivering 125bhp. Compared to the 36bhp in the 1.2-litre 1955 car, it's a positive dragster but, relatively, the 4/4 is gloriously unchanged in purpose.

The first 4/4 – four wheels and four cylinders – featured Morgan's sliding pillar/coil spring front suspension, and ash framing for the traditional-looking steel or light alloy bodies. Early 4/4s used Coventry Climax or Standard engines before Ford was adopted wholesale. The 4/4 was offered, successively, with Ford Anglia 100E sidevalve, Anglia 105E overhead-valve, 1.3-litre Consul Classic and 1.5-litre Cortina power before the long-standing 1.6-litre, 95bhp Kent engine was standardised in 1969. Later came Escort CVH and, briefly, Fiat and Rover twin-cam options. From 1969 a four-seater body was reintroduced, ideal for open-top family enjoyment. Front disc brakes, and four- and five-speed gearboxes were progressively introduced.

The Morgan 4/4 has unique post-vintage thoroughbred appeal with a rock-hard ride, flexing body and evergreen looks. Morgans aren't that expensive but they're great investments, a fluctuating waiting list making them virtually depreciation-proof.

MORGAN'S ROOTS

Henry Frederick Stanley Morgan, a vicar's son from Stoke Lacey, Hereford, began his professional career as an apprentice at the Great Western Railway works in Swindon and in 1909, by now a garage owner, 'HFS' created a three-wheeled light car to take an air-cooled vee-twin Peugeot engine. He designed and patented a sliding pillar front suspension system, and persuaded Mr Stephenson Peach, engineering master at nearby Malvern College, to machine its components for him. Friends were so impressed he decided to start making them commercially. After HFS scooped a gold medal in the Motor Cycle Club's 1911 London–Exeter–Land's End trial, the demonstrations of durability saw sales rise to 50 a week. Their popularity was down to amazing performance, keen pricing, and motorcycle-level road tax. Morgan made 15,000 three-wheelers up to 1952.

▲ *HFS in an early car.*

Facts & Figures: The Morgan Plus Four & Plus Eight

The Plus Four of 1950 was the first Morgan with a big engine, a 2.1-litre Standard Vanguard lump (later, 2-litre Triumph TR). The sloping radiator cowling style and faired-in headlights still used today arrived in 1955. When four-cylinder TR engines became obsolete, Morgan installed the 3.5-litre Rover V8 engine to create the Plus Eight in 1968. It was fabulously fast, with 0–100mph in 19 seconds and 125mph top speed – for the very brave.

◄ *Today's sportiest trad Morgan, the Roadster.*

▼ *The Morgan 4/4 can be had with either two or four seats.*

THE MORGAN AEROMAX

If ever you lamented the passing of can-do attitude in the British car industry, the Morgan Aeromax story will lift your flagging spirits.

When Swiss banker and Morgan racing addict Prince Eric Sturdza decided he couldn't find the perfect long-distance luxury sports car, he asked Morgan in late 2004: can you build it for me? His wishlist included air conditioning, a leather-lined cabin, cruise control, automatic transmission, and a flat load area with luggage access that wouldn't cause a hernia. Shaping his dream car would be 22-year-old Matthew Humphries, a recent car design graduate with abundant artistic talent.

His drawings created the most spectacular Morgan yet, with its 'gullwing' boot openings incorporated into the low-roofed fastback body. Once the prince approved them, Humphries built a quarter-scale clay model at the University of Coventry's Automotive Design School, all the time ensuring his design could be based on the standard Aero 8 – Morgan's 4.8-litre, 367bhp, V8-engined supercar, and its first model with an aluminium frame. In fact, a chassis beam was utilised by Humphries to stop the prince's baggage barreling into the two bucket seats in emergency braking.

The contours were digitised using special software. From this data, simple jigs were made on which the bespoke components were formed. This one-off Aeromax

The flowing lines of the Aeromax cloak a sports car expressly designed for long-distance touring.

The design, by a 22-year-old Matthew Humphries, incorporates boot openings on both sides.

then swiftly took shape. The body was hand-beaten over an ash wood frame, in time-honoured Morgan tradition. It wowed the Geneva motor show in March 2005, replete with a fully-fitted nest of Italian Schedoni luggage and Humphries' specially-designed door handles and wing mirrors.

People loved the Aeromax so much Morgan had to bow to pressure, with the prince's blessing, and build a limited edition series of 100 replicas costing £110,000 each. It sold out instantly, including to buyers Richard Hammond and Rowan Atkinson. 'As its driver, you feel you ought to be the hero of some sort of *Boy's Own* yarn', said a rapturous *EVO* magazine, after sampling one. 'Adventure should dog your every step,' they said, adding that the side-exiting exhausts give it: 'an exhaust note that makes Brian Blessed sound demure.'

Facts & Figures: The craft of Morgan

'The Morgan Aeromax demonstrates the wealth of talent in our small yet dedicated motor works,' said Charles Morgan at its launch. He is the grandson of the company's founder. 'It took just four months from a sketch to being able to test drive a real finished car,' he continued. 'The car is a showcase of the coachbuilding and technical skills of the whole team.' One lovely example of that is the laminated ash frame of the bodywork; it is incidentally beautiful, as each of the three backbones of the frame's skeleton has nine laminates of wood visible inside the car. Meanwhile, the wood above the dashboard and on the doors, solid ash, is carved to display age and grain.

THE MORGAN AERO SUPERSPORTS

The Morgan Motor Company is unique: still family-owned after more than 100 years. To celebrate this momentous achievement Morgan devised the Aero SuperSports. It has detachable aluminium roof panels to turn it from a coupé to a mobile suntrap, when they can be stowed in the boot.

THE MORRIS OXFORD 'BULLNOSE'

William Morris started a bicycle factory in Oxford and then ran a garage before building his own cars. He bought a disused military college in the Oxford suburb of Cowley, and in it he assembled his products almost entirely from bought-in components to keep investment costs down. The engine, for instance, was a four-cylinder 1,018cc sidevalve made by White & Poppe in the West Midlands, while Germany's Bosch supplied the magneto required for ignition. The leaf-spring suspension, the three-speed gearbox, the acetylene headlights, oil-powered tail lights and the braking system (for rear wheels only) were all delivered by suppliers, in contrast to the 'we-make-everything' attitude at Ford.

Another reason Henry Ford needed fear no disturbed sleep was the cramp-inducing nature of the new £165 'Morris-Oxford'. The short-wheelbase, pressed steel chassis wasn't big enough to accommodate four-seater bodywork, and the driving position had been created around William Morris and his agent Gordon Stewart, who were both vertically challenged gentlemen.

None of this deterred customers, who soon took the Oxford to their hearts as a trustworthy runabout. But William Morris was ambitious: in 1914 he visited the USA on a fact-finding mission, and was astonished to be offered engines from Continental in Detroit for just £18 apiece, against the £50 he was paying White & Poppe.

THE FIRST MORRIS MINOR

In the battle to dominate the British car market, it took Morris seven years to meet the wildly popular Austin Seven head-on. It finally did with the 847cc Morris Minor in 1929. It couldn't come a moment too soon, either, because sales of large-engined cars imploded after the Wall Street Crash that year, and the bleak economic winter that followed.

◄◄ The original Bullnose Morris, the Oxford, was craftily built from several suppliers' parts.

The Bullnose

The radiator grille surround of early Morrises sported a domed top, which made it look like an upturned 0.303 bullet. Hence the nickname 'bullet-nose', soon shortened to 'Bullnose'. It was either clever marketing or a happy concidence that the Morris emblem featured an ox crossing a ford, which was taken from the coat of arms of the city of Oxford. Still, sentimentality had no place in Morris's plans, and the distinctive Bullnose radiator was replaced by a conventional flat one in 1926.

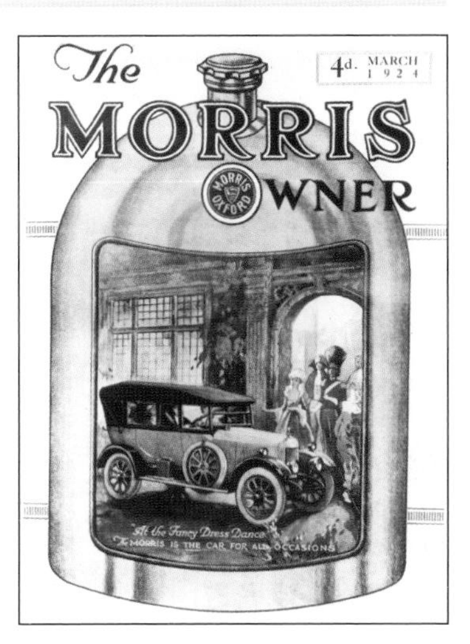

◄ The enlarged Morris Oxford from 1919 onwards proved an extremely popular family car.

Morris duly sent his top draughtsman, Hans Landstad, to Detroit, and during the voyage he designed a new Morris, the Cowley, which had a longer chassis, four seats and a 1.5-litre Continental motor.

Duties and sea battles (half Morris's 1915 consignment of Continental engines was lost when a cargo boat was torpedoed in the Atlantic) made importing components tricky. Morris eventually bought the rights to Continental's engine and got them made in Coventry.

In 1919, the Morris Oxford was back after a two-year gap in the form of an enlarged, upmarket edition of the Cowley. A longer wheelbase, more powerful engines and four-wheel brakes were progressively introduced. However, Morris's fortunes were changed by the 1920 Motor Car Act, which swung annual tax advantage in favour of the small-engined Oxford/Cowley at the expense of the large-engined Ford Model T; Morris finally grabbed British market leadership, and Bullnose sales soared.

THE MORRIS MINOR

In 1948, the new £358 Morris Minor was an economy car that drove, steered and handled outstandingly yet was still roomy, affordable and economical. Company founder William Morris, by then Lord Nuffield, was disdainful, stating the car resembled 'a poached egg'. However, the sour old duffer had to eat his words when the Morris Minor chalked up 1 million sales – the first British car ever to do that.

Its creator Alec Issigonis sought a new benchmark in roadholding. So he moved the engine further forward in the nose of the car for a better centre of gravity, specified rack-and-pinion steering, and designed torsion bar independent front suspension, which gave the Minor a less jarring ride than rivals. A throwback to older Morris cars was the 918cc sidevalve, four-cylinder, 27.5bhp engine, which meant the Minor huffed and puffed to its maximum speed of just 60mph.

Many early Minor MM-type saloons and convertibles were earmarked for export, and in 1950 the headlamps were moved from the grille to high up on the front wings, to satisfy American lighting rules. Ironically, transatlantic sales collapsed thereafter because Americans found the Minor too underpowered.

In 1950, too, a four-door saloon was announced, the two-door Traveller estate following in 1953, by which time the car had gained an 803cc overhead valve engine. In 1956, the Minor's engine size was boosted to 948cc to become the Morris Minor 1000. In 1962, engine size jumped again, to 1,098cc.

➤ *You still see Minors everywhere, but not many of these very early cars with their low-set headlights.*

A historic moment in several ways – the Minor was the first British car to pass the 1m manufacturing total.

Facts & Figures

Ideal everyday classic

As an everyday car, the Morris Minor makes sense. Unlike most classic cars, you can keep a Minor running indefinitely. Spares are cheap and plentiful, which is why Minors are still so ubiquitous. Charles Ware, who founded the Minor Centre in Bath in 1976, estimates 65,000 Morris Minors remain on British roads. 'We have really been responsible for the car's survival in Britain,' he said. 'They're emotive cars, people don't want to lose them. We can supply 99.5 per cent of the parts you need to build a Minor from scratch. The great thing is you can keep on using it while you do it up bit by bit.'

THE MORRIS MINOR TRAVELLER

Alec Issigonis devised the rear of the Traveller as a timber structure, cladding the frame with aluminium panels. There are 80 interlocking structural body parts to it. Along with van-like doors and flat glass, the rump of a Traveller fitted together like Lego . . . making it easy to restore today when trying to rid a Traveller of the inevitable woodworm.

The Minor lingered on until 1972 when the last, a Traveller, was made at Cowley, Oxfordshire. By the end, it was motoring's equivalent of a Chelsea Pensioner. Minors are sedate performers, but make up for that with excellent ride and steering, plus delightful details like the flashing green light on the end of the indicator stalk, the central speedo, and that comforting, farty noise when you change gear – caused not by any engine ailment but by the shape, or bore, of the exhaust.

THE MORRIS MARINA

Having long been the object of derision among car enthusiasts, the Marina was actually planned to be boringly conventional from the word go. This was the first all-new car created by the newly formed British Leyland, and it had one critical function: to lure customers away from Ford, and especially those in charge of buying fleets of company cars. These were clients who shunned technical wizardry for low prices and a wide range of engine, trim level and bodywork options with which rising employees could be appropriately rewarded.

That meant, in essence, copying the sort of cars Ford produced, and those tended to be rear-drive, reliable, boxy, roomy, and vaguely American in their styling. In these areas, the Marina hit the mark, but where the Morris fell down was in corner-cutting imposed by cash-poor British Leyland. The Marina's front suspension, for example, was a lever-arm set-up co-opted from the elderly Morris Minor, and its limitations endowed early 1.8-litre examples with bad understeer and disconcerting road behaviour in sudden manoeuvres. This was only partly fixed by a suspension modification. Smaller-engined models, using the faithful 1,275cc A-Series engine, weren't especially lively, and there were only ever four-speed gearboxes. They were all noisy, with a bouncy ride quality.

So, you could ask with justification, what's Britain got to be proud of here?

Well, by avoiding the front-wheel drive and fluid-filled suspensions found in other British Leyland cars, the Marina did gain a decent reputation for mechanical dependability and easy maintenance. Marinas sacrificed compactness to give plenty of metal-per-pound, and most owners found it provided a wholly reliable motoring experience. The Marina actually sold quite

Facts & Figures: The Morris Ital

Okay, we've cut the original Marina some slack, taking its obvious limitations into account, but the 1980–4 Morris Ital simply stretched the Marina's lifespan to its very fibres by way of a cheap-and-cheerful, £5m cosmetic facelift and an improved A-Plus Series engine that extended servicing intervals to 12,000 miles. For a time there was a 2-litre automatic range-topper too. TV commercials emphasised the involvement of Ital Design in revamping the car but, although Giorgetto Giugiaro's company had contributed some engineering advice, it was a bit of a con – the redesign itself was handled in dreary Birmingham, not sun-kissed Turin.

◄ *Marina 2 models, from 1978, were little improved from previous cars, but continued to find ready buyers.*

THE MORRIS MARINA VAN

Of the early Marinas, the least convincing was the TC, a two-door fastback barely able to cope with the 95bhp from the twin-carburettor MGB engine that could, just, propel it to 100mph. But the Marina's mechanical basic-ness, with live axle and rear leaf springs, was a positive boon in the van version, a workhorse that made a worthy equal to Ford's top-selling Escort cargo-hauler.

well, despite the destabilising strikes that so publicly brought British Leyland into disrepute. It was the UK's second best-selling car in 1973, passing the magic million sales post in 1978 soon after BL's new 1.7-litre O Series engines debuted in the Marina. The Marina never won any awards for anything, but as simple family transport it was actually one of British Leyland's better efforts.

THE NISSAN MICRA

The trusty Micra has been a mainstay of Nissan's commitment to this country, making its Sunderland plant (opened in 1986) Britain's biggest and most productive. It now makes a quarter of UK cars. The K11 Micra began rolling down the production lines in Washington in 1992.

This supermini newcomer boasted a rounded, almost retro, appearance, a pod-like instrument panel on a minimal dashboard, and refined fuel-injected 1- and 1.3-litre engines, the latter with 16 valves. The intention was to reel in trendy young buyers; it was critically acclaimed, becoming the first Japanese-badged car accorded European Car of the Year in 1993. However, as ever with the Micra, its attraction to the sensible 'grey pound' was magnetic, as controls were light, mechanicals reliable, and interior access easy. Sunderland struggled to oblige: a second assembly line was hastily installed, and by 1998 a million had been produced there.

An all-new Micra arrived in 2002. It was bolder, with headlights near the windscreen and irresistible, toy-like controls. It proved even more popular, with 250,000 sold in 20 months. UK Micras were exported to 44 countries. Sadly, the Micra's British era has now ended, with the 2010 car hailing from India. We'll miss it.

Facts & Figures

The Nissan Micra K10

The original Micra of 1983, always a Japanese import, became a massive seller in Britain for its almost tedious reliability and 'European' feel. The upright shape was boxy, it had three doors and a puny 1-litre engine (other configurations followed). Its durability was legendary: in 2006, official figures showed 30 per cent of the 340,000 K10 Micras sold here were still going strong.

▲ *The Micra has waved the Union Jack for the last time.*

➤ *The Angel of the North, plus a strange local sculpture.*

THE NISSAN QASHQAI

Britain is central to a revolution in compact family cars. Time was when, if that's the kind of car you needed, then you meekly bought a Focus, Astra or Golf. If you harboured a desire to drive a Range Rover then you kept it to yourself; after all, the appeal wasn't so much the frankly unnecessary four-wheel drive, more the practicality, high-riding driving position, and expensive, luxurious image.

The Qashqai (pronounced Cash-Kai – named after a nomadic Iranian desert tribe) that replaced the mediocre Nissan Almera in early 2007 is the answer to many a carbuyer's prayer. It's a 'crossover' fusing off-road stature with the tight dimensions of a conventional hatchback, while the cabin, complete with commanding seating positions, looks and feels decidedly upmarket. In most 1.6- and 2-litre models, only the front wheels are driven, as the majority of customers require, but 2-litres can also come with four-wheel drive.

It's proved astonishingly successful, with the UK, Russia and Italy the three biggest markets. A stretched seven-seater Qashqai+2 came along in 2008. The cars hit the half-million sales record in 2009, and 1,000 are built daily in Sunderland. One in six cars sold in the UK in 2009 was a Nissan Qashqai.

Facts & Figures: The country-wide connection

The Qashqai was conceived in Britain for pan-European markets. The car's athletic, big-wheeled design took shape in Nissan's London studios in a former British Rail maintenance depot in Paddington. It was translated into a real car at Nissan's European Technical Centre at Cranfield, Bedfordshire. Then it's manufactured at Nissan's super-productive Sunderland plant. It passed its Euro-NCAP crash tests with five stars and the highest score ever seen.

The Qashqai has been nothing less than a runaway success for Nissan and, above, its Sunderland plant.

THE NOBLE M12

Thanks to the M12 – and other sports cars in his bulging portfolio – Lee Noble has become something of a Godfather to the British sports car business. He's been at it since the early 1980s, usually adhering to the same racing car-inspired formula: extremely lightweight spaceframe chassis, mid-mounted engine, the most powerful engine he can find, and an emphasis on aerodynamics first and beauty second.

The M12, launched in 2000 to replace the unsuccessful M10, had all of this, but this time was reasonably affordable because Mr Noble farmed out much of the manufacturing to low-cost suppliers in South Africa.

All M12s came with a turbocharged Ford Duratec V6 powerplant, and performance of the glassfibre/composite-bodied two-seater was always spine-tingling. But not spine-jarring: the M12 had remarkably good ride, achieved by ditching anti-roll bars for firm settings in the double-wishbone suspension. So track handling was as enjoyable as everyday commuting.

The ultimate evolution of the M12 was the 3-litre GTO 3-R. *Autocar* magazine evaluated one in 2003. It found this 'rather wonderful little sports car' had 'superb traction everywhere and, on a circuit, it's an even more predictable car to slide.' America's *Road & Track*, meanwhile, found the 352bhp, twin-turbo M12 GTO-3R needed a mere 3.7 seconds to attain 60mph, could reach 170mph, and pulled 1.2 in lateral G-force during high-speed cornering.

Alas, the sports car industry is a precarious, internecine one. Lee Noble sold the company bearing his name, and by 2008 had left altogether. Rights to the M12 cars had already passed to American

The Noble M12 GTO 3-R can rocket to 60mph in 3.7 seconds, and in cornering can pull 1.2 in G-force.

THE FENIX

Lee Noble began again, from scratch, in 2009, founding Fenix Automotive to add a new sports car to his canon that so far includes the Ultima in 1983, the Ascari Ecosse in 1998 and, from 1999, the many Nobles themselves. The sports car world is on tenterhooks to discover how it measures up.

interests when the M14 was unveiled at the 2004 British motor show. Supposed to be a more civilised Noble to unsettle Porsche and Ferrari, it was soon sidelined by the all-new, 185mph M15. Now that one seems on permanent hold, which is unsurprising in the hostile economic environment. 'The M12 is a great car, but it's very focused and I wanted to produce a supercar people could use everyday,' said Lee Noble. 'It was time for Noble to take a big step up in refinement, practicality and style.' In the end, he's had to go it alone to try to achieve that.

The Noble M600

Noble continues without its founder, and throughout 2009/10 has been putting the finishing touches to the Noble M600. Departing radically from the old idea of affordable excitement, the M600 is pitched at the upper echelon of mid-engined sports-racing cars, where its twin-turbo Yamaha-Volvo V8 engine will offer 650bhp, a six-speed gearbox and an all-carbon-fibre body. Its backers like to highlight its 225mph capability . . . and that it can reach 100mph in 6 seconds. Fifty are envisaged a year, at £200,000 apiece. *Autocar* declared it had 'raw, brain-mangling power'; so, not a car for your granny, then.

◄◄ The focus was on driving enjoyment, but the M12 was always surprisingly affordable.

◄ The man himself, Lee Noble, has vowed to return with a brand new car, the Fenix.

THE PANTHER 6

Something highly unusual streaked across our fuzzy colour TV screens on Sunday afternoons in 1976: it was the Tyrrell P34, the world's only six-wheeled Formula 1 car. Meanwhile, Gerry Anderson's *Thunderbirds* could still be found on an endless loop of repeats on Saturday mornings. You could definitely see where Robert Jankel, Panther chief and a former children's clothes designer, got his inspiration for his Panther 6 – the talk of the 1977 Motorfair at Earl's Court, where it made an indelible impression on everyone who saw it.

His intention was to manufacture and market the world's first six-wheeled road car with four steerable wheels at the front, as well as aiming to offer the first standard street machine that could top 200mph. In a bitter blow to everyone who cheered for imaginative British automotive thinking, both goals were missed by miles. But at least he built two extraordinary, working (sort of) prototypes, and got everybody fired up to the idea that Britain could come up with a supercar every bit as astonishing as the monsters that rolled out of Turin.

One bulging pair of wheels at the back had 265/50 VR16 tyres, but the two pairs of steerable front wheels had 205/40 VR13 tyres. Allegedly, this was the bit that stymied the 6 because Pirelli, initially confident it could supply the right rubber, later backed away from the awesome responsibility of keeping the Panther firmly planted on the road. Which, with its twin-turbo Caddy V8 roaring away at the back – an 8.2-litre powerhouse – would have been essential.

It's all academic, of course. Yet had the Panther 6 ever become a proper showroom model, it would have offered a phenomenal specification, including a detachable hardtop, electrically adjustable seats, air-conditioning, an in-built fire extinguisher (slightly worrying, that), a telephone and even a standard television. Jankel guessed that few owners would try to experience the projected top speed: it would be enough for most just to *say* it could manage 200mph. . .

In the tail of this extraordinary roadster lurked a turbocharged 8.2-litre Cadillac V8.

THE PANTHER SOLO

Another sensational Panther that never really made it was the Solo. It was conceived in 1985 as an affordable mid-engined sports car, a latter-day Fiat X1/9, but the success of Toyota's MR2 put paid to that plan. Then it was transformed into a turbocharged, four-wheel drive supercar. The urge and road manners were exemplary but Panther simply ran out of cash to build more than a dozen examples in 1989/90.

Facts & Figures: The Panther Lima & Kallista

The Panther 6 was to remain a dream machine, but the quality and flair of Panther's craftsmen was widely available in the two-seater Lima, a 1930s-pastiche sports car introduced in 1976. It began life as a 2.3-litre Vauxhall Magnum, whose upper metalwork was replaced by a glassfibre roadster body with two-tone paintwork, wire wheels and a spare slung across the stubby tail. It made a fetching alternative to a Morgan. A tubular chassis frame was introduced in 1979 with the Mk II, and it was relaunched as the Kallista in 1982, now with an aluminium body and a choice of Ford engines.

The six-wheel steering system and front subframe laid bare.

Panther's craftsmen racing to get the car ready for Motorfair in October 1977.

THE RANGE ROVER

When revealed to an incredulous press on the Cornish coast in June 1970, there was nothing else like the Range Rover – nothing combining such tremendous, go-anywhere ability with luxury car comfort. The Range Rover had taken four years to perfect by a tight-knit Rover team led by engineers Spencer King and Gordon Bashford, and stylist David Bache.

Its rugged separate chassis supported permanent four-wheel drive for grip, coil-spring suspension for comfort, and Rover's aluminium V8 for effortless power. Clothing it was a body of new-suit crispness that, while utterly functional, managed to exude upper-crust style. Panels were of aluminium, with the notable exception of the drop-down steel tailgate, a notorious Rangie rust-spot.

Within a year, the Range Rover waiting list stretched into infinity. It scooped numerous awards and trophies, and remains the only vehicle ever exhibited at the Louvre as a piece of modern sculpture. Jeep, Toyota and others made large, four-wheel drive station wagons but, next to the Range Rover, they were as utilitarian as refuse trucks.

Not that the early Range Rover was exactly sumptuous. The vinyl seats and rubber floor mats were designed to withstand gun dogs' teeth and a good hosing-down. The design team was proud of the dashboard because it could double for right- and left-

From the bottom a very early Range Rover, a 1990s 'Classic' and the 1994 replacement codenamed Pegasus.

The timelessly crisp lines of the Range Rover are as noteworthy as its usefulness.

hand drive with little alteration. But, being made from coarse plastic, it looked rather primitive, and the radio was jammed into a slot between the steering column and the door.

It was a decade before the Range Rover saw significant change: a four-door version appeared in 1981, automatic transmission in 1983, and myriad trim upgrades and engine boosts until, by 1992 and the introduction of the 4.6-litre LSE with air suspension, the Range Rover was a veritable 4x4 gin palace.

By then, of course, it faced cheaper 'recreational' rivals. The market exploded but, ironically, none managed to topple the Range Rover's mix of off-road guts and feel-proud factor. The Range Rover became as potent a symbol of Britain as the classic red phone box, and the final old-shape Classic models came off the Solihull line in March 1996, after 317,615 had been built.

Facts & Figures

The blessed Range Rover

The Pope's visit to the UK in 1982 created a huge fuss. The government commissioned two Range Rover-based 'Popemobiles' for John Paul II so he could travel in majesty and safety, thanks to bullet-proof glass up to 'heart level' in the rear section where the pontiff stood (on thick blue carpet) and waved.

THREE GENERATIONS

Today's Range Rover is the marque's third incarnation. The original was joined in 1994 by what had been called 'Pegasus' while being honed by Land Rover engineers. It was a big step forward in technical detail, build strength and interior comfort, even if its styling was comparatively nondescript. There was also the option of BMW's excellent 2.5-litre, six-cylinder turbodiesel. Today's Rangie debuted in 2002, a totally new car with a unitary construction body and four-wheel independent air suspension. The car switched from its original BMW power units to Jaguar ones in 2006, making it a proper all-British car once more.

THE RANGE ROVER SPORT

Range Rover, once uncontested emperor of all off-roaders, has been under sustained pressure from European rivals. One sporty, premium 'sport-utility vehicle' after another has undermined its dominance. The first such upstart was the Mercedes-Benz M-Class in 1997, and that quickly elicited a response from BMW with its X5. Then Volvo (XC90), Lexus (RX400), Porsche (Cayenne) and Audi (Q7) piled in with ever more attractive, and ever less oversized, 4x4s.

Something had to be done. Land Rover's response was to rethink the Range Rover as the equivalent of a rough-terrain sports car. Revealed in 2005, it was only the fifth completely new model Land Rover had ever introduced. A design team led by Geoff Upex based the car on the Land Rover Discovery's semi-monocoque chassis, although its rakishly low roofline meant it would always be a five- rather than a seven-seater.

The whole concept was about shrinking the Range Rover around its core off-road capability. So there's a tight-fitting rear spoiler and sleeker rear lights for better aerodynamics, and a wheelbase 5.5in shorter than a Discovery's.

The Range Rover Sport arrived in supercharged form with a Jaguar-sourced 4.2-litre V8 petrol in 2005. It had a butch 390bhp and thumping torque of 410lb/ft; there was also a non-turbo 4.2-litre V8 and a 2.7-litre V6 turbodiesel. These were swiftly followed by a phenomenal V8 twin-turbodiesel. For 2010, the first two were uprated to 5- and 3-litre, both a third more powerful yet with less thirst and emissions. They all come with six-speed ZF automatic transmissions.

Land Rover was deluged by 20,000 orders for the Range Rover Sport, with its low roofline and short wheelbase.

◄◄ *The four-wheel drive car has immense on- and off-road performance.*

◄ *It's neat and purposeful from any angle.*

Facts & Figures: Ground-breaking, ground-hugging suspension

The Range Rover Sport's cross-linked air suspension allows for different ride heights for whatever terrain lies ahead. In one additional mode, it can individually adjust to traverse specific obstacles. The car also features Land Rover's Terrain Response system for conditions like grass, snow or sand, as well as overseeing traction control, hill descent control and several other driver aids. Plus, a dashboard display gives the driver a graphic warning if any wheel has absolutely no ground contact. The 2010 models feature the world-first predictive shock-absorption system called Damptronic Valve Technology, to further optimise ride and control. All of which will make the simple art of whittling a stick with a penknife truly relaxing by comparison.

THE RANGE STORMER

The 2004 Range Stormer was an appetite-whetter for the Sport. This low-slung, three-door show car, the first complete concept vehicle ever from Land Rover, made headlines with its skeletal seats, massive 22-inch alloy wheels and, best of all, split-folding gullwing doors.

The 5-litre Range Rover Sport has incredible on-road acceleration, reaching 60mph in under 6 seconds. But it also has seriously impressive ability in the rough. And that's even before you've revelled in the seductively luxurious interior. Little wonder 20,000 orders cascaded in, and *Top Gear* magazine anointed it its SUV of the Year in 2005. Its editor drolly commented: 'BMW and Porsche make world-beatingly great sports cars, one or two of which look a bit like SUVs. Land Rover has only ever made great SUVs, one of which now goes a lot like a BMW or a Porsche.'

THE RELIANT SCIMITAR GTE

As a name synonymous with the lowest form of British motoring life – the three-wheeler – Reliant always raises a smile. Strange to think then, that in 1968 the Tamworth factory in Staffordshire came up with a stylish trend-setter that inspired imitation worldwide: the Scimitar GTE.

It was the first 'Grand Touring Estate', its trendy rising waistline and wedge-like profile skilfully masking its capacious 36cu ft loadspace, accessible through a glass hatch. And the full four-seater GTE was the first car with individually folding rear seats, an arrangement copied for virtually all modern hatchbacks.

Styled by Tom Karen of Ogle Design (also responsible for – whisper it – the infamous Robin) like all Reliants the GTE's body was moulded in rust-resistant glassfibre, and carried on a strong steel box-frame chassis. Ford's beefy 3-litre V6 engine in such a light car meant the Scimitar was fast (120mph) and reliable, not to mention frugal: overdrive-equipped editions returned an excellent 27mpg.

Reliant was deluged with £1m of orders for the £1,998 Scimitar GTE. It was a handsome car that tugged the aspirational heartstrings. Sloane Rangers didn't exist in 1969, but for those who split their lives between Chelsea and somewhere in the country, the car was spot-on. Prince Philip gave a Scimitar GTE to Princess Anne as a birthday present in 1970 (she bought six more subsequently, and was famously booked for speeding in hers), which enhanced its social standing no end. It was more grown-up than a Mini Cooper but more affordable than a Jensen Interceptor.

◄ Until the GTE's feted debut you could buy a sports coupé or an estate, but not both at the same time.

THE TRIPLEX GTS

Triplex needed a showcase for its Sundym tinted glass, and asked Ogle Design to help. Designer Tom Karen did so by modifying a Scimitar GT into an unusual and attractive estate car with roof – plus side and rear windows – constructed totally of green-tinted Sundym glass. The Triplex GTS made its debut in 1965 to massive acclaim, with the Duke of Edinburgh particularly impressed. Quick-thinking Karen said he'd be welcome to have it on loan for as long as he liked, and Prince Philip accepted. From it sprung the inspiration for the Scimitar GTE.

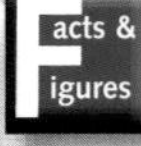

The original Reliant Scimitar

Reliant entered the luxury coupé arena in 1964 with the Scimitar SE4. This was a development of a unique vehicle, the Daimler V8-powered SX-250, built by Ogle Design. 'The fashion at the time was for cars to be "pulled out at the corners" and tend to have fins,' Karen recalled. 'The Ogle Daimler's "wraparound" design anticipated trends and so lasted much better.' The Scimitar itself switched to six-cylinder, 2.5- or 3-litre Ford engines. But once the trend-setting GTE arrived, the older coupé quickly faded from the scene.

Its ability to swallow loads made the Scimitar the ideal weekend getaway car for the Chelsea set.

In 1975, Reliant consolidated the success of the SE5 GTE with the wide-bodied SE6 but, somehow, this softer, more comfortable Scimitar tried to sneak its way into the executive car market. Ford's latest 2.8-litre V6 became standard from 1980. Cheaper, mass-market imitators bit into GTE sales and, by the early 1980s, Reliant couldn't afford to replace it. The last car was built in 1986, destined for one customer with an especially heavy heart – Princess Anne.

THE ROLLS-ROYCE SILVER GHOST

The 1907 Silver Ghost is probably the most famous British car ever, and possibly the most valuable anywhere, since it's insured for $57m. However, it last changed hands in 1948, when Rolls-Royce bought it back from a London owner who covered 500,000 miles in it over four decades of European motoring holidays.

This magnificent beast was the twelfth example built of Henry Royce's 40/50hp luxury car. It had a six-cylinder engine – still relatively novel – and an overdrive fourth speed (the 'sprinting gear') was hot stuff when most motorists avoided gearchanges unless absolutely necessary. The bare chassis impressed visitors to the 1906 Olympia Motor Exhibition and afterwards, by order of Rolls-Royce's managing director Claude Johnson, it was given the elegant silver 'Roi-des-Belges' touring coachwork and silver-plated brightwork that have made it so iconic.

Johnson nicknamed it 'Silver Ghost' and used it to demonstrate Rolls-Royce's durable quality and mechanical refinement. Soon all 40/50hps were called Silver Ghosts. Lawrence of Arabia praised his for helping seal victory in his desert campaign during the First World War; many 1914–18 Ghosts were built as armoured cars.

Facts & Figures

An epoch-making trial

In May 1907, the Silver Ghost undertook a 15,000-mile, six-week run observed by the RAC. Claude Johnson and the Hon. C.S. Rolls, with two colleagues, covered nearly 2,500 miles a week, running non-stop Monday to Saturday. They shuttled between London and Glasgow 27 times. Afterwards, the RAC stripped the car to measure wear: the cost of replacement parts amounted to under £3. The jaunt led *The Times* to declare the Ghost 'the best car in the world' – an epithet redolent of Rolls ever more.

Lawrence of Arabia with the Ghost he respected so highly.

Yes, this is the one – insured for $57m!

THE ROLLS-ROYCE PHANTOM III

Hailed as the ultimate luxury car, this Phantom III surged ahead again of its rivals, whose six-cylinder engines had become as smooth as Rolls-Royce's, and rather more responsive. Beginning in 1932, sixteen engineers spent two years designing the new V12-powered car. Their work was painstaking: the first prototype didn't run until June 1934, and the production Phantom IIIs didn't reach expectant buyers until May 1936. And this eternity was despite two decades' experience building V12 aircraft engines.

There was little true innovation. Other engines routinely used weight-saving light alloy, and Cadillac already offered a V12 (and even a V16) with hydraulic tappets to quieten the valves. Power output of 165bhp, moreover, was unremarkable for 7.3 litres. The engine's neatest feature was its double sparkplug-per-cylinder, each operated by separate ignition systems to banish lag between the two sparks.

Its sheer refinement, however, was breathtaking. The car was beautifully, astonishingly executed, and a good Phantom III today drives like a car 50 years its junior, with delightful steering and gearchanges.

The last of 727 Phantom III chassis was built in 1939, although not delivered to its owner until 1947. There wouldn't be another V12 Rolls until the Silver Seraph of 1998.

With imposing HJ Mulliner Sedanca de Ville coachwork.

Plans for a similar body from HR Owen.

Facts & Figures

A heavyweight contender

A Phantom III cost an eye-watering £1,900 in 1936. That was only for the chassis: customers needed to separately commission bodywork from a coachbuilder. Most were heavy limousine bodies, rendering performance slothful for a V12 engine, with an 85mph top speed and 0–60mph acceleration taking 17 seconds. No wonder these gigantic cars had a built-in jacking system! The most famous Phantom III was owned by the villainous Goldfinger in the eponymous 1964 James Bond movie.

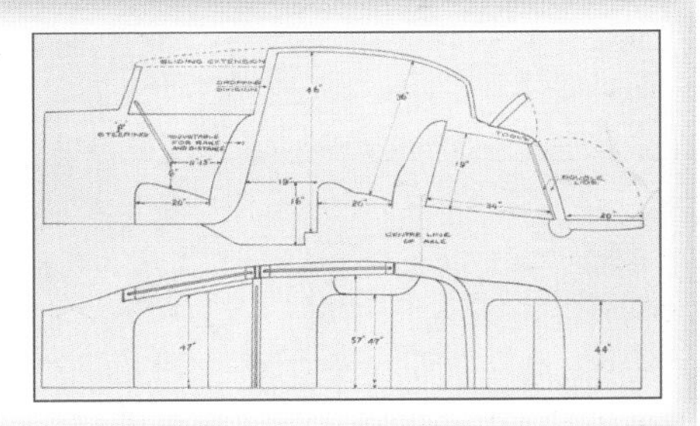

THE ROLLS-ROYCE SILVER CLOUD

John Blatchley, who died in 2008, was the genius stylist of the 1955 Silver Cloud. But, at the time, it was no fun.

'I spent years working on one, full-size mock-up of "the new car" only to be told when it was finished to put it on the bonfire,' remembered Blatchley. Rolls-Royce directors, as fiercely conservative as their clientele, viewed his work with a mixture of suspicion and horror. 'It was too modern. So I was asked to do a quick sketch of something more in keeping with the Rolls image, which I did in about 10 minutes. It was taken into a board meeting and they decided to make it.'

Orthodox or not, the result of that 10-minute sketch has stood the test of time. From the prow of the Palladian Rolls-Royce radiator grille, bestrode by the Spirit of Ecstasy mascot,

The Bentley S-type

A Bentley edition of the Silver Cloud was always available. Indeed, the original S-Type – perhaps indicating a slightly more relaxed, less formal British society – outsold the Cloud I; 3,072 Bentleys versus 2,238 Rollers. At the end of their lives, though, the tables had turned: there were 1,318 S3s against 2,297 Cloud IIIs. The principal differences were badges, radiator grilles and snob value, as they were built by the same people, in the same factory, at Crewe in Cheshire. Today, the plant is exclusively Bentley territory, yet busier than ever.

Who would have thought the design of the wonderful Silver Cloud was dashed off in 10 minutes, eh?

STANDARD STEEL

Only once the Bentley Mk VI/Rolls-Silver Dawn arrived in 1946 could a complete car, chassis and bodywork, be purchased from Rolls-Royce. The off-the-peg saloons soon gained the moniker of 'standard steel'. They accounted for 80 per cent of cars sold; you could still buy your Rolls or Bentley in chassis-only form to have it individually coachbuilt, but the enormous costs involved saw this indulgence gradually dwindle.

Series III Bentley and Rolls showing off their lowered bonnet line but resolutely different radiator grilles.

What's the bleeding time? Cloud owner James Robertson-Justice need only have consulted the dashboard clock.

to the tapering boot, the Silver Cloud is supremely elegant.

It rode on a heavyweight box-section chassis with independent front suspension and rear damper rates adjustable from the driving seat to suit different road conditions. From the comfort of that leather armchair, in front of the lustrous walnut dashboard, you could also lubricate the suspension. A quick prod on a pedal oiled it all to keep things floating along serenely.

The engine was the 4.9-litre Silver Dawn/Bentley R-type straight-six, with new aluminium cylinder head and twin SU carburettors. Transmissions were four-speed manual or Hydramatic automatic, manufactured under licence from General Motors, but after 18 months the unpopular manual option disappeared. It was a refined car but an unhurried performer, so the 6.2-litre V8 for the Cloud II in 1959 produced a welcome transformation. Surging mid-range acceleration and overtaking were now assisted by a likely 200bhp (Rolls never divulged precise power figures), and oodles more torque.

In 1962, the SIII introduced four headlamps and a lower bonnet line to slightly soften the stately image. John Blatchley was well aware of the Silver Cloud's special appeal: 'I came into designing Rolls-Royces thinking of them as flying drawing rooms. My one passion was to perpetuate that, to keep a rather nice idea intact.'

THE ROLLS-ROYCE SILVER SHADOW

It was a shock for traditionalists, the Silver Shadow. In one fell swoop, the flowing lines of the old Silver Cloud were redundant. Here was a square-cut, modern Rolls with features like self-levelling suspension and all-round disc brakes. It was even more silent and silken to travel in than its predecessor; faster, more economical, and more of an owner-driver's car at a time when Rolls-Royce chauffeurs were becoming the exception, not the norm.

The revolutionary aspect of the Rolls Shadow was its abandonment of a separate chassis. The monocoque (chassis/body combined) superstructure was steel, with aluminium doors, bonnet and bootlid. Aristocratic owners might not have twigged while admiring the lustrous paintwork and gleaming chrome, but this car was bang up-to-date.

Still, 'The Best Car in the World'? Some uncharitable souls dubbed it 'The Best American Car in the World', because the Shadow rode and handled similarly to Detroit's best luxury cars. The all-round independent suspension had hydraulic height control at both ends, a system supplied by Citroën. The power steering was vague and featherlight. If you hurled the Shadow into corners, it would wallow, understeer and roll alarmingly compared to, say, the Mercedes-Benz 300SEL 6.3. Crewe engineers, of course, would have been appalled at such loutish driving techniques. One contemporary report said: 'First and foremost, the Shadow is a superlative journey car, for driver as well as passengers.'

The 6.2-litre V8 engine and GM Hydramatic four-speed automatic gearbox were standard, carried over from the Cloud. This elegant magic-carpet-on-wheels was endowed with a 115mph top speed, a 0–60mph

◄ The Silver Shadow II had many worthwhile improvements like split-level air-conditioning.

➤ *Who wouldn't like to waft through Piccadilly Circus in the cosseting comfort of a Shadow?*

THE BENTLEY T-SERIES

There was a Bentley version of the Shadow, called the T-Series or, retrospectively, the T1. Although differences were minor and cosmetic, limited to 'B'-winged radiator grille, badges and small décor items, Bentley owners saved £64 in 1965 by not opting for Rolls cachet! However, Bentley was then in severe decline – only 8 per cent of Shadow I/T1s were Bentleys, compared to 40 per cent of the previous Cloud/S-Series.

Facts & Figures: A different kind of coachbuilding

Without a separate frame, one-off cars such as had been created on the Silver Cloud chassis by many independent bodybuilders weren't feasible any more. So Rolls filled the yawning chasm in the range for more 'personal' models, the two-door Silver Shadow Coupé and Convertible, by handcrafting them individually in-house at its subsidiary Mulliner Park Ward. They were distinguished by an undulating waistline profile. Renamed Corniche in 1971, they would remain part of the Rolls line-up until 1987.

Stirling Moss always knew a winner when he saw one, racing car or not.

time of 10.9 seconds, and a standing quarter-mile covered in 17.6 seconds. Fuel consumption was good by deep-pocketed Rolls standards: you could expect between 12 and 15mpg on a run, and maybe 8mpg around town.

Changes came slowly until the engine was enlarged to 6.75 litres in 1970: three-speed torque converter transmission replaced the fluid flywheel system in 1968; 1969 saw a long-wheelbase version, when air-conditioning and a safer, US-style dashboard was adopted. The Shadow II of 1977 gained modernities like a front airdam, split-level air-con and rack-and-pinion steering. These Rollers sold better than any previous models; 32,038 by the time the Silver Spirit replaced them in 1980.

THE ROLLS-ROYCE PHANTOM DROPHEAD COUPÉ

You've got to hand it to the current Rolls-Royce regime: in a world where anything new is scrutinised to the nth degree by rulemakers and safety experts, they've managed to pack the Phantom Drophead Coupé with features that cannot fail to delight the fortunate owner.

The most extravagant aspect of the swish four-seater convertible are its huge, front-opening 'coach doors' that swing out to facilitate an entrance direct from a 1930s Hollywood spectacular. No other car on the road has these. They're made possible by the strong aluminium spaceframe within, which makes the car lighter than it looks and extremely rigid; along with taut suspension settings, sharpened steering, and a Sport mode in the automatic transmission, this sybaritic motor car is the most responsive Rolls ever.

The Drophead was previewed by a concept, 100EX, exhibited at motor shows in 2004. From that showpiece the brushed steel bonnet and A-pillars made it to production. They're machine-finished and then hand-polished for a flawless sheen. The teak decking for the rear hood cover is also derived from 100EX, treated with blended oils to preserve its natural finish. Meanwhile, the boot lid splits into two, the lower tailgate dropping down as a picnic seat.

The fabric roof is the largest on any modern convertible. Its five layers, including a cashmere lining, drown the hubbub of the outside world, and when raised, at night, it's peppered with pinpricks of light like a starry sky – a sensational effect delivered via dimmable fibre-optics.

➤ *The Phantom Drophead began life as this concept sketch for 100EX.*

➤➤ *The 'coach' doors are made possible by a very strong aluminium structure.*

The first five Dropheads left Rolls-Royce's factory in July 2007, bound for new owners abroad. After what seemed like decades of stagnation and formality at the 'old' Rolls-Royce, this was a stirring moment. The style of this new breed of Rollers, masterminded by BMW, divides opinion. But the commitment to traditional materials and handcrafts – the Drophead Coupé has 1,300 new parts over the Phantom – is a shot-in-the-arm for prestige British design and manufacturing.

THE GERMAN CARVE-UP

Defence firm Vickers, Rolls-Royce Motors' owners since 1980, decided in 1998 to sell its hallowed carmaker. In the ensuing auction, Volkswagen beat bitter rival BMW to snatch it for £430m, unaware that rights to use the Rolls-Royce name weren't included! These rested with Rolls-Royce plc, and the firm preferred to bestow them elsewhere, to their aero engine partner . . . BMW! Volkswagen got to keep Bentley and the production plant in Crewe, and the two marques were formally separated on 1 January 2003 after 72 years.

➤ *Teak decking on the hood cover is treated with blended oils; it reflects the boatbuilding heritage surrounding Rolls-Royce's Chichester factory.*

Facts & Figures: The Rolls-Royce Phantom

The Drophead Coupé is built – by hand – alongside the Phantom saloon, the four-door limousine that relaunched Rolls-Royce in 2003 from its new assembly plant at Goodwood, West Sussex. Both Phantoms employ an alloy spaceframe construction method, which entails up to 140 metres of hand-welding. They feature a V12 engine with the long-established Rolls capacity of 6.75 litres. Its performance is described as 'brisk': top speed is limited to 155mph, with 0–60mph taking, for the Drophead, 6.1 seconds, and 5.7 for the saloon. Joining them at Goodwood is the Ghost, a smaller, more compact saloon, and the hardtop Phantom Coupé.

THE ROVER 3.5-LITRE P5B

Margaret Thatcher arrived in Downing Street in 1979 to begin her 11-year reign. The country quickly felt the effect of all the changes she had in mind for the way Britain was run. But one thing that wouldn't change for a few years yet was the mainstay of the Downing Street car pool, the Rover 3.5-litre.

They had been obsolete for a full six years but these stately Rovers had performed an understated job of serving Prime Ministers Callaghan, Heath and Wilson before her on their official business around Whitehall, shuttling to and from Heathrow, and ferrying these leaders around Britain throughout the turbulent 1970s. Harold Wilson even had a special ashtray fitted to one to cope with the regular emptying of his trademark pipe.

Nor was Maggie the only influential woman to enjoy the Rover's well-bolstered comfort and creamy power from its Buick-derived 3.5-litre V8 engine. Her Majesty the Queen, no less, was a devoted owner, regularly seen driving hers around Windsor. With customers like these, Rover found itself in the exalted position of making the car the Establishment favoured.

It had been introduced in 1967, the injection of American-style horsepower rejuvenating the Rover 3-litre P5 – a fine, dignified car but, until then, always a ponderous one. In the transition to becoming a 184bhp, 110mph motorway express, the delightful African cherrywood, leather and Wilton-carpeted interior was maintained, and you could still opt for the unusual Coupé bodystyle, offering a lower roofline while keeping four doors (a style recently revived and

Not just any old 3.5-litre, this is the actual car often seen around Windsor with Her Majesty at the wheel.

THE ROVER 3500 SD1

British Leyland pulled a decent car out of the hat in 1976 after a parade of duffers. The 3500's fastback styling was the height of modernity, and under that sleek, Ferrari-inspired bonnet was Rover's familiar V8, mated to a brand new five-speed gearbox. The crushed velour interior, complete with dials contained in a prominent box atop the dashboard, was impressive. The euphoria was crowned by the big Rover's elevation to European Car of the Year for 1976. However, in its eight years on sale, the 3500 suffered patchy build quality that allowed BMW, and even Saab, to outshine it. The British crest was fallen. . .

The Rover P4

The dignified, dependable P4 series typified Rover's approach to cars throughout the 1950s. A car for the professional middle and upper-middle classes, it was discreet, comfortable and fastidiously constructed, as any barrister, doctor or city gent would require. As the 60, 75, 80, 90, 95, 100, 105 and 110, over 14 years it came with a galaxy of four- and six-cylinder engines, and plenty of detail differences between each version.

copied by Mercedes-Benz and Volkswagen). But the car now stood out on its racy Rostyle wheels.

These Rovers sold well until 1973, when the ageing 3.5-litre was pensioned-off. No other 1970s British saloon possessed the same aura of solid worth and confident sobriety, and the Jaguars that ultimately usurped them in the Thatcherite 1980s were altogether more spivvy.

THE ROVER P6

Rover invented the 'executive' car with the 2000, first in its P6 line, a vehicle as mould-breaking in its class as the Mini was among economy cars. This 100mph saloon was ripe for Britain's unfolding motorway age, and didn't appeal solely – like previous Rovers – to dour, silver-haired bank managers in bowler hats. It was ultra-modern, while never appearing trashy.

The 2000 bristled with new technology, majoring on safety long before it was fashionable. It featured a super-strong 'base unit' (a steel skeleton to which unstressed outer panels were attached), four-wheel disc brakes, and a well-padded interior. Indeed, the chances of avoiding a crash were increased because the racing car-style De Dion rear suspension meant the 2000 had prodigious road adhesion, and could well handle the flexible power of its overhead-camshaft, twin-carburettor engine. American safety campaigner Ralph Nader hailed it as 'how all cars should be built.'

Then Rover truly excelled itself by installing its light-alloy V8 engine, to create the Three Thousand Five. This made it a rapid and covetable Jaguar-chaser, with surprisingly good fuel economy and, in Mk2 3500S form with a manual gearbox, very fast – a favourite 1970s police pursuit car.

The Rover T4

Looks familiar? You wouldn't have thought so in autumn 1961 when Rover revealed this space-age four-door saloon prototype. The main focus was underneath, where a gas turbine engine powered the T4 via the front wheels. But two years later the futuristic T4 style emerged to gasps of surprise and delight, little altered (but without a jet engine, of course), in Rover showrooms as the 2000.

The Rover P6 was applauded for its focus on safety; this one was owned by Dunlop, along with the A1 registration number.

THE ROVER JET 1

The crowds who flocked to the Festival of Britain exhibition on London's South Bank in 1951 may have sensed the nation's motor industry was on top of its game. On display was Rover's Jet 1 –the first ever gas turbine-powered car.

The Jet 1 project team, led by Spencer King and Frank Bell, overcame numerous engineering obstacles. The 230bhp engine turned at 26,000rpm, five times higher than most normal car engines in the early 1950s; it was air-cooled, and turbines have no engine braking. But Rover really made it work. In 1952, Jet 1 established a turbine car speed record of 151.96mph. That must have gladdened the heart of jet engine inventor Sir Frank Whittle.

The terrible fuel economy of turbines – Jet 1 managed just 5mpg – and the expense of building the motors precluded them from use in everyday cars (Chrysler got closest, loaning 50 experimental cars to the public for real-world evaluation). Jet 1 was the first of many prototypes that gained massive media coverage so, for Britain, the Rover Jet 1 was still a prestige asset.

Rover's ground-breaking gas turbine car proving it really did take to the road.

Facts & Figures

The Rover-BRM

Rover's adventures with roadgoing gas turbines cannot pass without mention of its Le Mans escapades. The BRM-built, Rover turbine-powered two-seater gained special permission to run in the 1962 24-hour race (although not in competition). Graham Hill drove it to complete the course, covering 2,592 miles at an average of 108mph. Two years later, a redesigned Rover-BRM, piloted by Hill and Jackie Stewart, was an actual competitor, and that too finished the race – the first British car home – averaging 99.8mph.

THE ROVER 75

It used to be that when a British car 'headed east', it was enjoying some sort of limited success in Japan. Alas, for the rather splendid Rover 75, that phrase now has a more ominous ring.

After five years as an independent, British-owned company, MG Rover collapsed in 2005, making the venerable Longbridge factory and its 5000-plus workforce redundant. It was very sad, especially as Rover was 101 years old and the plant itself was marking its centenary. Yet when all the huffing, puffing, recrimination and hand-wringing was finished, it became a clear-up operation – an asset sale to whoever offered the best prices. Sentiment counted for nought.

So the Rover 75 was sold off to Chinese manufacturer Nanjing Auto, and everything associated with it (bar the Rover name, which Ford had first claim on) was shipped to Shanghai, to be reborn as the Roewe 750.

THE ROVER 75 4.6 V8

In one of several inventive ways to keep Rover going while a partner was (fruitlessly) sought to share future cost burdens, the company attempted to resurrect the golden era of the Rover 3.5-litre (see page 160). They re-engineered the 75 in 2004 with rear-wheel drive and a powerful 4.6-litre Ford Mustang V8 upfront. In its MG ZT incarnation, the car was now a real muscle-machine, but in Rover form, more a high-performance cruiser. A little under 900 were built before the official receivers moved in.

Under BMW, Rover's reputation for sophisticated luxury cars was briefly restored, with the attractive 75 saloon.

Facts & Figures

Great car, bad atmosphere

No sooner had the Rover 75 been unveiled to an appreciative media throng at the 1998 British Motor Show in Birmingham, than it was dogged by controversy. Rover chief Bernd Pischetsrieder stood up and lambasted the Government for dragging its feet over subsidies to renovate Longbridge; he hinted darkly that drastic action might be necessary. This was a terrible shadow to cast over the 75, the pride of the entire Rover workforce whose morale he had just crumpled. Fortunately, the British public had the sense to recognise a genuinely good car, and the 75 won many friends.

Loving attention to design detail produced these chrome-rimmed white dials.

Those lucky old Chinese car-buyers can now enjoy the fine qualities of this executive car, with its soothing, front-drive platform and suspension honed to perfection under BMW's exacting eye (BMW owned Rover from 1994 until 2000), its marvellous interior with veneered panels and chrome-rimmed instruments, its range of aluminium K Series engines from a 1.8-litre four-cylinder to 2.5 V6, and one of the first five-speed automatic gearboxes.

The 75 was conceived for the more mature driver who valued refined comfort over hard-riding, hollow-feeling, plasticky modernity. This Rover would be an executive saloon that dodged overt sportiness for comfort, smoothness, discreet style, and something almost entirely absent from the sector – old-fashioned charm. A miniature Bentley, if you will, equipped to swan benignly through the mêlée of Britain's overcrowded roads. Some said it was too retro but it picked up copious awards, including 'world's most beautiful car' as adjudged by a panel of Italian aesthetes. Until that calamitous day in 2005, it steadily rebuilt Rover's status as a maker of fine British motor cars.

THE SS 100 JAGUAR

Partners William Lyons and William Walmsley lit up the motorcycling world in 1922 by proving sidecars needn't look like bath chairs. Their wind-cheating Swallow sidecars made a motorbike combination look a 'pillion dollars'. They were soon sprinkling their magic on car bodies as well and, by 1927, offered stylish coachwork on popular chassis from Austin, Standard and Swift.

Swallow Sidecars prospered, moved from Blackpool to Coventry, changed its name to SS Cars, and Lyons bought out Walmsley to assume sole control. He wanted to be a car manufacturer proper and, in 1931, took the plunge by launching the raffish SS1.

Although it oozed Sunset Boulevard glamour, the SS1 sports saloon used ordinary Standard components in a special chassis. Consequently, Lyons' cars offered unbeatable style for the money. Other models soon followed. But few set the pulse racing like the SS 100 of 1936.

A low-slung two-seater sports car with long, elegant mudguards, the SS 100 looked ultra-desirable. It was the first car to bear the 'Jaguar' name and elegant leaping Jaguar radiator mascot, sculpted by leading motoring artist F. Gordon Crosby.

Beneath its overly long bonnet was an SS-reworked overhead-valve version of Standard's seven-bearing 2,663cc straight-six, producing a healthy 104bhp and driving through a four-speed gearbox. The SS 100's chassis

The SS 100 evokes, for many people, the very essence of the low-slung 1930s sports car; it was a genuine 100mph machine to boot.

was textbook 1930s, with an 8ft 8in wheelbase, beam axles front and back, and underslung rear suspension.

It was a sexy-looking car that could also hold its own on the road, pulling 80mph in 2.5-litre form, while a much-tuned SS 100 lapped Brooklands at 95mph. 'Up to about 60mph, the engine is scarcely noticed, except for increasing exhaust note,' said *The Autocar* in September 1938.

Then a 125bhp version was launched in 1938, with a 3,486cc engine, enabling the roadster to hit the magic 100mph. This SS 100 embarked on two years of successful rallying, and its last major victory was in the Alpine Rally as late as 1948.

The SS 100 itself wasn't revived after the Second World War (the final example was delivered in 1941, the last of 314). But one thing did roar back into life: the by-now unfortunately entitled SS Cars quickly changed its name to . . . Jaguar.

Facts & Figures: The one and only SS 100 coupé

This gorgeous two-seater fixed-head coupé version of the 3.5-litre SS 100 was revealed at the 1938 London Motor Show at Earl's Court. Although announced as a £595 production car – a standard open one, widely considered one of the best-looking SS Jaguars anyway, cost £445 – it remained a one-off, effectively rendering it the first Jaguar concept car. It was bought by the wealthy parents of one Gordon March, the lucky recipient of the car as a 17th birthday present.

THE PANTHER J72

With imitation being the sincerest form of flattery, Panther paid the SS 100 the ultimate compliment in 1972 by using the car as inspiration for the J72. This was one of the earliest 'replicars': they are generally loathed by purists, but the J72 at least used Jaguar engines and running gear, and the handmade quality of its chassis, aluminium coachwork and beautifully-trimmed cockpit was obvious.

THE STANDARD VANGUARD

Nationalisation fever gripped Britain after the Second World War. First the railways and then the coal industry came under state control. Car company chiefs feared they'd be next, so Standard's Sir John Black tried to secure his place in any new order: Standard would make the tough, roomy, mid-sized but – most critically – export-orientated car its *métier*. That's how the Vanguard, named after battleship HMS *Vanguard*, was conceived at breakneck speed in a febrile political atmosphere.

To a cross-braced Standard 14hp chassis was fitted an all-new 2,088cc four-cylinder engine with 'wet' cylinder liners to ease replacement; a robust and eager unit. The full-width body with six side-windows resembled a streamlined Hudson shrunk in the wash. Indeed, Standard's stylist Walter Belgrove had been dispatched to the American Embassy in London for inspiration, to sketch all the latest American sedans he saw.

The car industry was commanded by the new Labour government to 'Export or Die'. Standard duly sent the Vanguard abroad by the boatload, sometimes as kits for local assembly of saloons, estates, vans and pick-ups. The nationalisation threat evaporated. Meanwhile, the Vanguard's hasty development showed through. Prototypes had been tested only in Britain, and production cars struggled with hot weather and rough roads in far-flung corners of Blighty's remaining colonial markets. One problem was dust filling the interior because the doors had no rubber seals; another that the 65mph Vanguard shook itself to bits on cobbled surfaces like Belgium's punishing *pavé*.

The Vanguard adopted the streamlined American look in a bold attempt to conquer world export markets.

The Vanguard was modified as each nark was revealed. Overdrive arrived in 1950, a bigger rear window in 1952. Standard made the car lighter by (no, really!) using thinner metal for its body panels. The 'Phase II' of 1953 was roomier but the distinctive, slope-back rear end was changed to a conventional 'notchback' tail.

Back in 1947, though, the few Brits who could get hold of a Vanguard – you were only entitled by national need, which tended to favour doctors, farmers or anyone else crucial to the community or economy – you were proud to own your country's newest and most forward-looking motor car.

BRITAIN'S FIRST OIL-BURNER

The Phase II Vanguard brought something genuinely novel to British motorists: a diesel engine in a private car. The 2-litre engine came from the Ferguson tractor, and the Vanguard was reinforced to take its extra weight. It was a rough, clattering, slow old thing. The payback, though, was a potential 50mpg which some industrial fleet customers, like the Port Talbot steelworks, truly valued.

Facts & Figures: The Standard Vanguard Phase III

In 1955 came a totally new style of Vanguard: the Phase III. It was faster, more economical, lighter and, with a roofline cut by 3.5in, a lot more svelte. Presently, there was a small-engined one, the Ensign, mostly sold to company fleets and the RAF, and the Vanguard Luxury Six, fitted with an all-new 2-litre straight-six engine. Standard could not shrug off the drabness of its own name that was, well, a bit standard-sounding, even though it was supposed to evoke patriotic flag-hoisting. Standard was axed altogether in 1963, its place taken by the altogether more rousing Triumph.

THE SUNBEAM ALPINE

The stylish Alpine, for years, was assailed with jeers about being a car for ladies or hairdressers. This probably reflected the unenlightened era when it was deemed perfectly okay for paunchy, balding inadequates to demean women (and colourists). Because the Alpine, never the fastest sports car on sale between 1959 and 1967, was still one of the most civilised and thoughtfully designed.

The Rootes Group endeavoured to come up with something to clobber MG and Triumph in export markets, primarily North America. Perhaps that's why a prominent tailfin was included either side of the capacious boot – trendy, for sure, but these faddish touches added to enthusiast scorn.

The Alpine was practical, with two tiny back seats for toddlers (although bags of shopping fitted better), wind-up windows, and a decent canvas roof. Its 78bhp, 1.5-litre four-cylinder engine was, essentially, like any you'd find in a Hillman Minx, albeit with an aluminium cylinder head. Acceleration was leisurely, 0–60mph taking 18.8 seconds, but the Alpine could screech to a standstill aided by front discs. 'Although it cannot be described as hairy,' said *Cars Illustrated* magazine in January 1960, 'it offers a first-class performance for a 1.5-litre production sports car.'

Alpines underwent constant updates. The Mk II of 1960 had an extra-flexible, 80bhp 1.6-litre motor, and

Potted history of Sunbeam

Founded in 1899; emerged as Britain's premier name in 1920s grand prix racing; made fast, high quality sporting cars – ideal if you couldn't afford a Bentley; part of a three-way Anglo-French merger in 1925 that never gelled; taken over by Rootes Group in 1935; melded to form Sunbeam-Talbot in 1937, which then turned out attractive sports saloons using Hillman and Humber components; two-seater Alpine of 1953 bore the Sunbeam solo name again, acquiring some rallying glory; Rapier followed in '55, Alpine in '59; RIP: 1976.

◄ *Later cars, from the Mk IV onwards, were toned down stylistically, with their fins banished, but all Alpines were handsome.*

tougher suspension. The Mk III of 1963 gained a revised suspension and servo-assisted brakes, while twin fuel tanks, one in each wing, doubled the boot volume; a Mk III GT was offered with a detachable hardtop and no folding roof mechanism.

The following year brought the Mk IV, a cosmetic rethink this time as those controversial fins were eliminated. Did it look better? Very debatable: you suddenly realised how much character the Alpine had had with them! Finally, in 1965, the Mk V delivered a 92bhp 1,725cc engine. Even with this punch the Alpine could never break the ton. Then again, the 100mph-plus MGB and the Triumph TR4 had bigger engines, so fair dos. Neither competitor was as pleasant or comfortable in daily use – the Alpine was always a first-class package . . . whether you worked in a salon or maintained oil rigs.

A ladies' car? A poodle was hairier than an Alpine, some said. . .

A roomy, comfortable cockpit, with proper wind-up windows, were rarely found on other British sports cars of the 1960s.

That's right – it's Cary Grant, in a Tiger.

THE SUNBEAM TIGER

Sunbeam had a wildcard crack at the American market by turning the meek Alpine into the fearsome Tiger. With advice from Carroll 'Mr Cobra' Shelby, a 164bhp 4.2-litre Ford V8 was shoehorned in, turning the car into a formidable 120mph trouncer of the Austin-Healey 3000. No sooner was the Tiger winning startled admiration in 1964 than Chrysler bought control of the Rootes Group. Disapproving of its Ford engine, the charismatic Tiger was poached after three years and 7,000 cars.

THE TRIUMPH ROADSTER

The Triumph Roadster oozes charm. The massive chrome headlamps, the long bonnet, the flowing wings and the 'dickey seat' – the last production car with one, making the Roadster a five-seater – combine to make it stand out from its many and interesting contemporaries.

The car was the brainchild of Sir John Black in his attempt to steal sales from Jaguar. In the '30s Black had owned a racy, two-seater SS 100 Jaguar, while a Triumph trademark in those days – particularly on its Dolomite roadster – had been a dickey seat. Black thought combining the two cars' elements made a winning formula.

The 1,776cc, four-cylinder engine, independent front suspension, back axle and four-speed gearbox all came direct from the firm's Standard 14 saloon. But to ensure the car had the right sort of sporting overtones, the power unit was the special, overhead-valve edition, refined by famed engine designer Harry Weslake.

The car's splendid coachwork was the work of one of Standard's draughtsmen, Frank Callerby, acting on Black's precise instructions. The rakish style would become familiar to millions from its appearance in TV detective series *Bergerac*; that long bonnet with the radiator grille set back over the line of the front axle, the three windscreen wipers, the closely hunched rear end, and the prominent, teardrop headlamps with their domed glasses. The boot lid doubled as a screen for the two rear passengers.

Priced at under £1,000 road-ready, the Triumph Roadster offered nimble handling and a firm ride. But it

The Roadster was one of the first new British sporting cars after the Second World War – check out the triple windscreen wipers and huge headlights.

It was a snug car with the hood raised . . . and you could get two people in the dickey seat in the 'boot'.

was a real slowcoach. It scored an epically tardy 34.4 seconds in *The Autocar* magazine's 0–60mph test, while top speed never exceeded 75mph.

In 1948, after 2,501 1800 Roadsters had been made, the 2000 Roadster was launched with the Standard Vanguard 2.1-litre engine and three-speed gearbox. This transplant greatly improved acceleration from 0–60mph, to 27.9 seconds.

The car was discontinued in 1949, a proposal for a streamlined, gadget-laden successor never getting beyond prototype stage.

Sir John Black

Black, managing director since 1934, was behind the success of Standard and Triumph cars after 1945. His personal vision created diamonds, like the Triumph TR2, and duffers, such as the Triumph Mayflower. But most of the firm's post-war income came from manufacturing Ferguson tractors, a typically canny Black deal. His style was aggressive and scheming, and fellow directors finally ousted him in a late-night putsch in 1954.

THE TRADITIONAL APPROACH . . . OR WAS IT?

Underneath the voluptuous skin of the Triumph Roadster lies a story of ingenuity in difficult times. Its tubular steel chassis had its origins in Standard's Second World War work building aircraft fuselages. The old-fashioned coachbuilding practice of aluminium panels on a wood frame avoided expensive tooling costs and tricky negotiations with the Government for steel supplies. Using pre-war Standard components like the gearbox and the back axle required precious little extra engineering. This all allowed Triumph to launch a glamorous new model in 1946, a full two years ahead of most of its peers. The car offered style over performance, but there wasn't much else around to upstage it.

THE TRIUMPH HERALD

In 1959, the fins and chrome of American cars reached their acme; Cadillacs looked like sci-fi rocket ships, and even run-of-the-mill saloons had gaudy jukebox dashboards. British cars had no such styling excesses. But we did have the uncommonly snappy Triumph Herald.

This new Triumph's smart suit of clothes was styled by an Italian consultant, Giovanni Michelotti. It was a dramatic departure from the boring old Standard 10 it replaced, perhaps no surprise as Signor Michelotti had penned quite a few Ferraris. You got chrome-hooded headlamps and small fins at the back as part of a neat package that was as sharp as an origami swordfish. There were saloons, coupés, convertibles and estates.

Technologically, the new car was a backward step. Standard-Triumph couldn't afford to tool-up for a mass-production monocoque body/chassis, so it reverted to an old-style separate chassis. The 948cc Standard 10 engine had to suffice, too, plus all-round drum brakes. However, compensations included a 25ft turning circle, rack-and-pinion steering and the first all-independent suspension on a small British car.

The Herald was on sale until 1971, by which time a separate-chassis car was an almost prehistoric idea. But thanks to progressively bigger engines, the Herald always felt competent. And it was always chic.

How many Rolls drivers really did take the Herald instead, one wonders?

Vitesse identifiable by its quad headlights.

Facts & Figures

The Triumph Vitesse

Thanks to the Herald's Meccano-like construction, other models could be created from the basic platform. The Vitesse, launched in 1962, was unusual in packing a small-capacity six-cylinder engine, at first a 1.6- but later a 2-litre, but always with twin carburettors and a close-ratio gearbox. The Vitesse was easily recognisable by its quadruple headlights.

THE TRIUMPH SPITFIRE

The Spitfire was a two-seater roadster version of the Triumph Herald, a budget sports car with lovely, undulating lines drawn again by Giovanni Michelotti. It inherited the Herald's swing-axle independent rear suspension. So enthusiastically driven Spitfires could be twitchy cars in ill-judged corners, when the suspension would jack them up at the back and unseat them. Only the reckless would have encountered serious trouble, mind.

With the wind in your hair and a rasping exhaust note, the Spitfire was intoxicating even on a jaunt down to Fine Fare. And it cost just £729. It eagerly outpaced the MG Midget, and was good for 90mph with 0–60mph in 16.5 seconds. Front disc brakes, a decent soft-top and roomy boot made it eminently practical.

The 1965 Mk2 gained 4bhp from its new 1,296cc engine, and there was 8bhp more for the 1967 Mk3, so the Spitfire could now hit the ton (preferably on a dead-straight road). In 1970, the MkIV offered elegant bodywork and 'swing-spring' rear suspension to cure the previously wayward tendencies. The final change came in 1975 when a 1,500cc engine meeting American anti-smog rules was standardised, and this Spitfire 1500 remained current until August 1980.

➤ Spitfires line up before shipment to happy customers in the USA, where it was a British sports car mainstay throughout the 1970s.

Facts & Figures

The Triumph GT6

The GT6 was a Spitfire with a fastback body, incorporating a lift-up tailgate, and a 2-litre, six-cylinder engine. The MkI was absurdly over-powered, the feeble Herald rear suspension not up to the 95hp delivered to the rear wheels. Later modifications tamed the beast, and the last ones built in 1973 were akin to a pint-sized E-type or DB6. Well, almost.

TRIUMPH TR6

Bit of a bloke's sports car, the TR6 – not a curve in sight, a ride like it's bouncing off moguls on a ski run, and a deep exhaust note likely to scare pets and annoy neighbours. Do-gooders like ergonomists and safety monitors clearly never went anywhere near the cockpit, as the focus here is on driver enjoyment. With hood raised, water and draughts wangle their way in, leaving behind a lingering smell of old-dog damp to intermingle with the odour of hot oil (gearbox) and sticky vinyl (upholstery). Great stuff!

ITALIAN BODY,
GERMAN FACELIFT

Weird, really, that the image of the TR6, a quintessentially British roadster, should embody so much foreign influence. The TR4/5 was styled by the Italian design consultant Giovanni Michelotti, and always recognisable by the quizzical look at its headlights. The lines were then spruced up by Germany's Karmann, as part of its contract to supply the TR6 tooling. They made a handsome fist of it, so Britain needn't have felt miffed. . .

If the TR6 didn't exist, you'd wish someone had invented it. But this classic British roadster did exist for seven glorious years between 1969 and 1976. Then, instead of cherishing and preserving it so there'd always be a TR6-style machine available for chaps after seeking a bit of raucous, open-top, wheel-spinning and tail-out fun, British Leyland dunderheads replaced it with the appalling TR7, aimed at girls.

They claimed they had to move with the times. The TR6 was old hat, able to trace its roots straight back to the

Never the most refined roadster you could buy, the joyous TR6 proved extraordinarily desirable nonetheless.

Triumph TR2 and, like that car, featured a separate chassis. Through TR3 and TR4, the concept of powerful, rough-and-ready fun was maintained, with Triumph finally incorporating independent rear suspension in 1964. With the TR5 three years later, however, came a powerful six-cylinder engine, turning it into Britain's first production car with fuel-injection.

The TR6 was simply a tidied-up version of this very car, with – in Britain – an identical 150bhp 2.5-litre motor for ramping it to 60mph in 8.8 seconds. Wider wheels, a front anti-roll bar, servo-assisted brakes and radial tyres were all welcome, nay necessary, attributes; 120mph was achievable, with flat cap wedged tightly on. In 1973, worthwhile improvements included a re-profiled camshaft – it was some 15 per cent less powerful overall, but much smoother – and a subtle chin spoiler. Most buyers opted for overdrive to make the four-speed gearbox less tiresome.

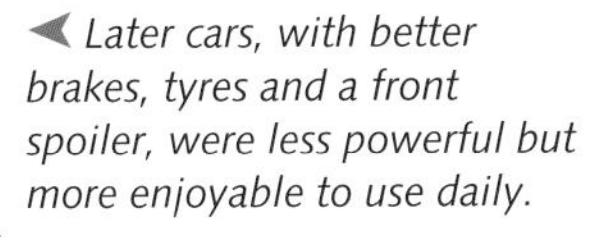

Later cars, with better brakes, tyres and a front spoiler, were less powerful but more enjoyable to use daily.

Facts & Figures: The damned Americans

America first exerted its influence on Triumph sports cars in 1967. The TR5's fuel-injected engine was too polluting for US emissions laws, and so the States instead received the near-identical TR250, with a twin-carburettor engine that, at 109bhp, was about a third down on horsepower. A similar situation arose for the TR6, where a hugely detuned 105bhp motor powered North American market cars, blunting performance and image in the face of new rival the Datsun 240Z, whose engine had no problems meeting US smog rules. All of this did matter, though: of the 94,619 TR6s built, only 8,370 were sold in Britain, most of the balance being bound for the USA.

THE TRIUMPH DOLOMITE SPRINT

Triumph had BMW and its 2002 in its sights when it devised the original Dolomite, a classic front-engine/rear-drive saloon melding compact style with invigorating performance; a proper sports saloon to take over where the well-liked, if flawed, Vitesse tailed off.

For the right calibre of four-door saloon with unquestionable 100mph credentials, engineers raided the corporate stores for the best bits from their other cars – all centred around a slant-four, overhead-camshaft 1.85-litre engine derived from a motor Triumph previously supplied exclusively for the Saab 99. The rear-wheel drivetrain came from the Toledo while the body was a smartened-up 1500 with four headlamps. It was loaded with admirable goodies like full carpets, a heated rear window, cigar lighter and a clock, while optional overdrive spelled more relaxed motorway cruising.

Yet, cursedly, BMW still had the edge on Triumph, with its exhilarating, fuel-injected 2002Tii. So a year after the fine-handling, 91bhp Dolomite made its debut, Triumph served up this riposte. The Dolomite Sprint featured an enlarged 2-litre engine, still single overhead-camshaft in design but with a 16-valve head. It delivered its meaty 127bhp through a TR6 four-

TRIUMPH

TANGLED '70s WEB

Triumph's 1300 saloon was unusual for 1965 in featuring front-wheel drive. Unfortunately, it sold poorly and was expensive to make, so the range was split in 1970, with an enlarged 1500 going upmarket and the cheaper Toledo – now with rear-wheel drive – aimed at the masses. In '72 came the rear-drive Dolomite. The 1500 then also went rear-drive to match it in '73, before the entire family of cars was renamed Dolomite in 1976.

Triumph's clever 16-valve engineering turned the Dolomite into a BMW-basher; well, almost. . .

speed-plus-overdrive gearbox. It had real alloy wheels, not the usual go-faster fakes of the time – the first British car so-equipped as standard – sat on lowered suspension, and looked pin-sharp with its black vinyl roof, twin exhaust tailpipes and purposeful chin spoiler.

The performance was superb by contemporary benchmarks, with 0–60mph acceleration taking just 8.4 seconds, and a 119mph top speed. You could order a limited slip differential to improve the already tidy road manners. But one area where the British newcomer had the BMW 2002Tii comprehensively in its shadow was value. Starting at £1,740, the Sprint was about £1,000 cheaper than the Beemer.

It won genuine acclaim, exemplified by *Motor* magazine which declared: 'Performance is there in plenty, and the model's manners quite impeccable. It is a tremendously satisfying car to drive.'

Rotten labour relations, plummeting build quality and a crumbling image for anything produced by British Leyland – Triumphs included – seriously hampered the Dolomite Sprint's chances of competing with its German foes long-term, which was a great shame.

The Sprint flew the flag for British Leyland in motor sport, but the corporation itself was nothing to shout about.

Facts & Figures

'Britain leads the way!'

This was the declaration from *Motor* magazine in 1973, as it described the then-new Dolomite Sprint. It could justifiably lay claim to being not just Britain's but the world's first mass-production four-cylinder car with a multi-valve engine. The British Design Council accorded Triumph an innovation award in 1974, and 16-valve engines have since become commonplace. So have compact sports saloons from Audi, BMW and Mercedes-Benz.

THE TVR M SERIES

The 'M' stood for Martin, Martin Lilley, at the TVR helm alongside his father Arthur since 1965. Father and son methodically transformed the company from a chaotic builder of superior back-garden specials into a pillar of the British sports car establishment.

Much remained from previous TVRs. The long nose/short tail look, for instance, and the snug cockpit defined by a gigantic transmission tunnel along the middle. And the poised, neutral cornering that made all TVRs such unadulterated fun to chuck around, abetted by the engine set behind the front wheels for a low centre of gravity.

But the Lilleys went through the whole car with a fine-tooth comb. The backbone chassis was strengthened and made easier to build, and a longer bonnet allowed the spare wheel to find a new home under it. The menu of power units included a four-cylinder 1.6-litre Ford or a 2.5-litre Triumph straight-six, and Ford's lazy, 138bhp 3-litre V6 from the Capri.

CLEVER TREVOR

Founded in a Blackpool backstreet by apprenticed mechanic Trevor Wilkinson, the TVR trademark is a contraction of his Christian name. Trevor was also an amateur car builder, creating 'specials' out of any parts he had lying around, while paying the rent by repairing fairground rides! He sensed he was on to a winning business when some of his cars were ordered from the USA to be entered in sports car races there. By 1958 TVR was a proper car company, although Trevor left in 1962 and had little more to do with TVRs as they increasingly offered credible rivals to Lotus. He died in 2008 in Menorca, after enjoying his retirement thereabouts on his small yacht.

This is the convertible S version of the TVR M Series, introduced in 1978, that offered a raucous driving experience and fine handling. Only 258 of these were built.

▲ *These 3000s kept TVR's Blackpool factory busy throughout the 1970s; they edged slowly upmarket but kept their raw sporting appeal.*

After 1973, the V6 3000M became the mainstay. The alliance of lazy Essex power and lively Lancashire sports car nous was well matched, a 121mph coupé with an 8 second 0–60mph time that became increasingly 'executive' in its character, but still with a seat-of-the-pants ride quality through that short-travel, ground-hugging suspension.

The Lilleys launched a 230bhp Turbo in 1975 – the first such car from any British manufacturer – and incorporated a practical all-glass hatchback in the Taimar of 1976. Finally, in 1978, there was the 3000S, a traditional open roadster which was rattling good fun.

Many felt something special was lost when the rounded, old-fashioned TVRs were swept aside in 1980 to make way for the controversial, wedge-shaped Tasmin cars. On paper, they were better in every department, but the 3000M's gutsy, hot-rod image was lost . . . albeit not irretrievably (see side story).

Facts & Figures

TVR S and 350i: back on form

After Peter Wheeler bought TVR in 1981, he rethought what a real TVR should be. His conclusion was the existing Tasmin wasn't the right man for the job, and so turned it into an awesome performance car by slotting in Rover's aluminium V8 engine. The insipid, droopy-nosed Tasmin now became a 140mph slingshot with a 0–60mph time of under 6 seconds. Then Wheeler turned his attention to a cheaper, entry-level TVR, and came up with the S. It was in the same spirit of the 3000S of 1978, only vastly improved under its curvaceous lines.

TVR CHIMAERA

The 1990s TVR everyone craved was the Chimaera. Little surprise, really, when the 4-litre V8-powered two-seater could hurtle to 100mph in about 12 seconds from a standing start, passing 60mph at 4.7 seconds and not running out of steam until the speedo showed 152mph.

TVR's traditional backbone chassis easily handled the performance, and the all-round independent wishbone suspension, and coil-over-gas dampers, enabled superbly entertaining handling and excellent cornering. There was a notable absence of electronic driver aids. The person behind the wheel had to get the best from the car, and power applied at the wrong moment could see the Chimaera easily into a spin.

However, in contrast to some other TVRs, here was a civilised car, a touring machine to use day in,

GRIFFITH AND FRIENDS

The beautiful two-seater Griffith heralded TVR's renaissance in 1991. It was the fearsome, hard-edged alternative to the Chimaera. Later, in 1995, came the Cerbera, a four-seater TVR which also saw the arrival of TVR's very own engines, the AJP8 and AJP6. These powered a succession of increasingly radical and extrovert sports car, including the Tuscan Speed Six, the Tamora, the T350 and the Sagaris. But in 2004, Peter Wheeler sold the company to the 24-year-old son of a Russian oligarch for a reputed £15 million. Either his timing was impeccable or else Nikolai Smolenski was inept but, by 2006, the company collapsed.

Facts & Figures: Ned's toothy trademark

TVRs were created by a tight-knit 1990s team, working with devoted company owner Peter Wheeler. Prototypes were built quickly and exhibited at motor shows – if the public liked them, the cars would soon be on sale. When the Chimaera was being designed, Wheeler's gun dog Ned was said to have bitten a chunk out of a foam styling mock-up. Everyone liked the shape this created so much it was incorporated into the finished car, as the recess for the front indicators!

day out, with a comfortable, luxuriously trimmed cockpit, and a particularly spacious boot. To that end, the suspension was tuned for a slightly softer ride. The ingenious and easily used soft-top just begged to be folded down – all the better to hear the spirit-lifting woofle from the exhaust pipes.

The Chimaera, named after a monster from Greek legend, was British to its core, its engines being Rover V8s progressively developed by TVR so you could at various times pick 4-, 4.3-, 4.5- and 5-litre versions; the gearbox switched early on from Rover to Borg Warner, and you could specify power steering.

As a compromise between the hard-edged power of other TVRs, like the Griffith, and the clinical but luxurious nature of a BMW or Porsche, the Chimaera made a wonderful choice. Surprisingly refined, too. Many successful city types, bonuses burning holes in their suit pockets, ordered one as a weekend fun car to blow away the worries of the pressured working week. Although sometimes bemoaned as a 'softer' TVR, it proved resoundingly popular, with 5,432 sold – indeed, the best-selling TVR ever. *Autocar* summed it up in a 4.5-litre Chimaera road test: 'Big grunt. Yet it's all so user-friendly. You can use nearly all of it nearly all of the time.'

THE VAUXHALL PRINCE HENRY

Kaiser Wilhelm II, the self-styled Prussian emperor, was at the peak of his powers as German ruler in 1906. His car-mad kid brother Heinrich, meanwhile, was roaring around Baden-Baden, Regensburg and Nuremberg, in a loop starting and finishing in Munich, driving a Benz 40hp on the yearly Herkomer Trophy.

The week-long competition required drivers to set numerous timed average speeds over its many sections. Cheating was rife. Heinrich loved the driving but detested the foul play. So in 1908 he decided to sponsor his own, annual 'Prinz Heinrich Fahrt' (Prince Heinrich Trial) with each driver accompanied by an independent observer.

His name attached huge cachet to the new annual event and attracted several factory-backed entries – including, in 1910, one from Britain's Vauxhall. Chief engineer Laurence Pomeroy believed competition success best proved the mettle of Vauxhalls, and so entered a team of three short-chassis, 3-litre A-type cars, driven by himself and colleagues Percy Kidner and A.J. Hancock. They weren't racing cars: Prince Heinrich planned the 1,200-mile trial as a punishing test for standard touring models.

They were timed at an impressive 70mph on the trial's long, rugged route, and all three finished,

➤ *The Prince Henrys – this is a C-type from 1911 – were formidable performers, and distinctive cars with their fluted bonnet sides.*

Facts & Figures

Fast, fearless, and great in bends

A revised Prince Henry was launched in 1912 with its engine uprated to 4 litres, giving it dazzling performance – with an ability to cruise at 65mph (the British speed limit was then 20mph!) – to match its outstanding road manners. 'It is less as if it cornered well than as if there were no corner there,' wrote the late motoring historian Kent Karslake after riding in the 1914 Prince Henry owned in the 1950s by the son of its designer, Laurence Pomeroy Jnr. Its competition record was awe-inspiring. In 1913, for instance, Prince Henrys won 35 hillclimbs, 23 Brooklands races, and 14 reliability trials.

THE VAUXHALL 30/98

The Vauxhall Prince Henry became the 30/98 in 1913 and 600 of these superb sporting cars found eager owners up until 1927 (happily, 170 survive). For a time it was sold with a written guarantee that it could do 100mph! This reputation for high performance probably attracted General Motors, which bought Vauxhall in 1925. Still, that marvellous heritage didn't stop it, subsequently, from turning the Luton-based company into a mass-producer of family cars.

a vindication of their inherent stamina. *The Autocar* magazine interviewed Kidner afterwards: how were the roads? 'Well,' he replied, 'in Prussia they're so bad our brick highways are like billiard tables by comparison!'

In 1911, Pomeroy unveiled his C-Type 20hp, and Vauxhall adopted the anglicised 'Prince Henry' name for the sports model. It was Britain's first proper sports car. Despite its uncomplicated, sidevalve engine, it devoured the road with gusto. Kidner drove a Vauxhall Prince Henry in the 1912 Swedish Trials. When temperatures sank to -30°C, the intrepid crew were compelled to survive on chicken thawed out and cooked on the hot exhaust pipe!

Around 190 Prince Henrys were made until the outbreak of the First World War. Then, Vauxhall built 2000 D-type staff cars for the British Army – many helping, ironically, to thwart Prince Henry's brother Kaiser Bill.

THE VAUXHALL CRESTA PA

Autumn 1957 had a decidedly stratospheric feel. The world's biggest radio telescope went into operation at Jodrell Bank, while Russia launched its first space satellite, shortly followed by Sputnik II, containing Laika, the first dog in orbit.

Vauxhall's new big saloons, the basic Velox and upmarket Cresta, seemed positively space-age themselves. Vauxhall set out, in 1955, to encapsulate some American-style glamour for the duo, and the effect was modern Detroit yet on a scale more in tune with Droitwich. The new cars had sleek, elongated lines, fashionable fins topping off the rear wings, bold rear light clusters and gleaming chrome trim framing the low-set radiator grille and headlights.

The remarkable windscreen wrapped around the sides of the car for a panoramic aspect, while at the back a three-

Facts & Figures: The 'mighty' Cresta Friary estate

Basingstoke's Friary Motors offered an eye-catching, commodious and practical estate car conversion of the PA cars from 1960. Only a tiny number were sold, and today they're very collectable, with about 40 known survivors. One unlikely owner was Queen Elizabeth II. Hers was delivered to Windsor in 1962 and she was still enjoying using it regularly until the 1970s. It bears the registration number MYT 1, and is stored at Sandringham.

◄ The Cresta brought colour and style to an often drab 1950s Britain.

► The PA's wraparound windscreen was a talking point.

piece rear window provided outstanding all-round visibility. With column-mounted gearchange and full-width bench seats front and back, six adults were accommodated comfortably.

The 2.2-litre six-cylinder engine was familiar from previous Vauxhalls. This power unit, mated to a three-speed manual gearbox, meant the car hit 60mph in 16 seconds and had an unspectacular 89mph top speed. Still, steering was light and roadholding acceptable, despite a tendency to wallow in corners.

These handsome vehicles brought colour and style to Britain's dreary roadscape. There was a raft of striking two-tone paintjobs, or you could order a vivid single colour like salmon pink or apple green. The Cresta sported whitewall tyres and anodised hubcaps like its American Chevrolet cousins, with upholstery in two-tone leather, herringbone-weave Nylon, or hardwearing 'Elastofab'.

In late 1958, a Cresta cost £1,073, including £358 Purchase Tax. It was the kind of car you desperately hoped your dad would get next. With a youthful Cliff Richard breaking into the charts and coffee bars springing up everywhere, these snazzy Vauxhalls were an integral part of modern Britain as the 1960s dawned.

A wraparound single-piece rear window to match the front one was new for 1960, along with an engine boost to 2.6 litres, and optional overdrive, giving 96mph top speed. The two-millionth Vauxhall, built in February 1959, was actually a Cresta, just one of the 173,604 PAs made. Perhaps it was never quite as gripping as the 'space race', but the PA Cresta genuinely enlivened British motoring.

THE CRESTA'S RUN

The Cresta name first appeared in 1954, on a fancy version of the Vauxhall Velox E Series intended to rival the Ford Zodiac in the flash stakes. Then, after the PA's demise in 1962 came the more restrained PB and PC models, later given a 3.3-litre straight-six motor. The last Cresta PC was sold in 1972.

THE VAUXHALL VIVA

The popularity of smart little saloons like the Anglia and Herald finally nudged Vauxhall into attempting its first economy car. Coincidentally, these two were built on Merseyside, where most Vivas would eventually hail from.

Vauxhall intended its new small car to slot in below the Victor in its range. The Viva had austere 'razor-edge' lines but a spacious four-seater interior. The 1,057cc, four-cylinder engine put out 44bhp, drive was to the rear wheels via a four-speed all-synchromesh gearbox, and steering was rack-and-pinion. It proved an easy-to-drive machine, with light controls and nippy handling.

THE ELLESMERE PORT STORY

Vauxhall was forced to build its Viva plant near an unemployment blackspot. It picked Merseyside, where the decline of the docks meant jobs were badly needed, and settled on an old airstrip between the Manchester Ship Canal and the River Mersey. Work began in 1961 on Vauxhall's Ellesmere Port factory. On 1 June 1964, it rolled out its very first Viva. In just 10 months, 100,000 were sold, and by the time production of this HA version halted in 1966, 307,738 had been built. This was followed by 556,752 Vauxhall Viva HBs up to 1970, and 700,000 HC cars throughout the 1970s.

Bringing a Viva home for the missus – it was marketed at women drivers in particular.

At first, in 1963, there was a choice of Viva or Viva Deluxe two-door saloons, starting at £527 including Purchase Tax. You could specify disc brakes at the front (for 12 quid extra), and it was the first British car to wear acrylic lacquer paint. Vauxhall principally targeted women buyers, another minor innovation.

In 1964 the range expanded to include a luxury SL, with polished radiator grille, and also the pseudo-sporty SL90, tuned to give 60bhp for 0–60 acceleration in 18 seconds and raise top speed to over 80mph.

The Viva HB in 1966 was longer, lower, wider, roomier and even more successful. Underneath its trendy lines were new coil-spring suspension, plus a gutsier (1,159cc) engine. The range soon included a four-door saloon, an estate, automatic transmission and a 1,600cc engine. In 1970 the HC Viva sported Americanised, barrel-sided bodywork adding to elbowroom inside, on a longer wheelbase. In addition to 1.2- and 1.6-litre engines, Vauxhall squeezed in 70bhp 1.8-litre and even 110bhp 2.3-litre units.

The Viva lingers in people's affections for its dependable, easy-to-live-with nature. Unfortunately, in real life, few Vivas linger; this minor British icon was more susceptible to rust than most.

Sporty HB Vivas

Australian former F1 star Jack Brabham endorsed a special edition of the 1.2-litre Viva HB in 1967, mimicking Cooper's tie-up with the Mini. The Viva Brabham delivered an additional 9bhp courtesy of twin carbs and a trick exhaust system, along with front disc brakes and a distinctive go-faster stripe draped over bonnet and front wings. The Viva GT went further still, with a 2-litre motor from the Victor under a louvred, matt-black bonnet – tilted at 45° to squeeze it in! This transplant spelled 104bhp and maximum torque of 117lb/ft; it was a peppy performer – 0–60mph in 11.9 seconds and 100mph tops – and *Autocar* loved it: 'When real acceleration is needed, the engine will pull like a train.'

THE VAUXHALL ASTRA

Once, if you insisted on buying a straightforward, British-made car, the default choice was a Ford or Rover. But neither of these venerable makes count any longer: Rover went down the tubes in 2005 and Ford just gave up, deciding henceforth all its wares would be imported. So now, if you want a mid-market, British-made car wearing a traditional British livery, the Vauxhall Astra is all that's left.

Fortunately, it's pretty good. The Astra has just entered its sixth generation as a class front-runner. This stratum of car has long been exemplified by the Volkswagen Golf, but since the fifth-generation Astra's quantum leap forwards in dynamic design and build quality in 2004, the compact Vauxhall's been snapping at its German rival's heels.

Today's Astra, unveiled at the 2009 Frankfurt motor show, was largely designed by a Brit, Mark Adams. On a wider track and longer wheelbase, Mr Adams and his team created a bold-looking car, whose 'cab-forward' stance incorporates lots of wing- and blade-shaped touches hinting at its artful streamlining – this Astra spent 600 hours being fine-tuned in a wind tunnel. Inside, a wraparound dashboard, groaning with buttons, creates an intimate atmosphere.

Engineers have rethought the Astra too, redesigning the rear axle and perfecting an optional Flexride

➤ The wing and blade influence in the styling of the all-new Vauxhall Astra is more than just decoration; its aerodynamics took 600 wind tunnel hours to fine tune.

BRITAIN LOVES ITS ASTRAS

According to official statistics, almost one in twenty, or about 5 per cent, of vehicles registered on British roads is a Vauxhall Astra. The Astra also accounts for a third of Vauxhall's sales, so it's a real benchmark for the British motor industry: the country's best-selling British-built car. We snap up one in every four Astras sold across the whole of the Eurozone, so the British view of the Astra is absolutely critical to General Motors Europe. Over 3 million Astras have been sold, most of them built at Ellesmere Port. The Mk I was an import until manufacture started in Britain on 16 November 1981. After that, the Mk 2 of 1984 featured super-aerodynamic design, the 1991 Mk 3 was more spacious and comfortable, and the Mk 4 of 1998 had massively improved quality and rust-resistance.

system so the driver can vary the damper settings for ride quality from standard to sport or touring. There's a wide engine choice, one of the best being a feisty 1.4-litre turbocharged petrol, although the 1.7- and 2.0-litre diesels are the stronger sellers. Astras also have bi-xenon headlights that adapt to the prevailing road and weather conditions.

A Vauxhall Astra will never be the most impressive car to namedrop, but it's in the uppermost percentile for all-round satisfaction.

Facts & Figures: The Vauxhall Astra Mk I

We'd never seen a Vauxhall quite like it; it was the first with front-wheel drive and a transverse engine, and made the Viva it replaced look positively Victorian. In 1982, the Astra also became the first Vauxhall to offer a diesel engine, while the sporty GTE version added another innovation to the Astra's tally: it was the first car sold in Britain with a totally colour-keyed body, featuring wheels, bumpers and mirrors all in the same bright white or red. . .

➤ *Is this your car, sir? More than likely, as some 5 per cent of cars on British roads are Vauxhall Astras.*

THE WOLSELEY 6/80

The 1948 Wolseley 6/80 became a favourite with British police forces for its excellent power-to-weight ratio, big drum brakes and a powerful 72bhp engine making this an 85mph car.

Wolseley supplied them in a heavy-duty specification to the Metropolitan Police, with beefier anti-roll bar and electrical systems. Wireless Area cars, for responding to 999 calls, sported a roof-mounted radio aerial and a chrome Winkworth gong in place of one front foglight; Motor Traffic Patrol cars, in addition, gained two Tannoy loudhailers on the roof, wider tyres, and illuminated 'POLICE' signs. Both had Pye radio transmitter/receiver units in the boot.

But the only thing making the 6/80 especially responsive was police drivers' own defensive driving techniques. In this context, the Wolseley 6/80 was as good a car as any for its role. Only after hoods started using stolen Jaguars as getaway cars did forces select their pursuit cars more cannily.

Nuffield's finest

Current from 1948 to 1954, the 6/80 was flagship of the Nuffield Organisation's range. It shared its body with the four-cylinder Wolseley 4/50 but with a bonnet extended by 7in to accommodate its 2.2-litre overhead-camshaft, straight-six engine. The comfortable interior included, unusually for the time, a standard heater. Downsides were vague and heavy steering, no temperature gauge, and an engine design fault that meant exhaust valves often burnt out after just 10,000 miles; many a police mechanic, with a 6/80 engine in pieces, had reason to curse Nuffield's misguided choice of weak metals for valves and guides! The 6/80 was also positively archaic in the level of regular greasing and oiling its steering and suspension required.

The 6/80 was powerful and spacious.

Battle-scarred 6/80s on the police training skid pan at Hendon.